Praise for *Feathers*

"A readable introduction to feathers and what they mean for birds and mankind." —*Science*

"[*Feathers*] is gracious, funny, persuasive, and wide ranging."
—Amanda Katz, *New York Times*

"[D]elightful. . . . [A] fascinating inquiry into one of those common things that are easy to overlook until someone shows what a miracle it is." —*Seattle Times*

"[Hanson] gives unexpected substance to his nearly weightless subject."
—*Discover*

"[C]aptivating. . . . Beginning with the evolution of birds, Hanson . . . unfolds the human fascination with feathers in terms of science, commerce, tools, folklore, art, and aerodynamics with panache."
—*Audubon*

"There are many feathery gems to be discovered in this book. . . . *Feathers* is a book designed to be read, not skimmed, with every page revealing new layers of understanding." —*Examiner*

"To read *Feathers* is to meet up with an enthusiastic old friend who simply cannot wait to tell you about something he just discovered. . . . a must-read by bird watchers and naturalists of all levels of interest or experience." —*Bird Watcher's Digest*

"His narrative excursions are so charming, you will hardly be aware of how much you're learning while you read. And the next bird you see—wow!" —*The Reader's Review*

"[A] delight." —*Montana Outdoors*

"The science, evolution, and practice of studying and identifying feathers are the subject of this delightful book for the general reader. The author traveled widely, interviewing specialists, to piece together what we know about feathers: the almost impossible genius of their many uses and effectiveness, the history of their study and use as ornament, and the current theories about their evolution."

—*Book News*

"This one is a great read!"

—Wayne Mones, *The Perch*,
Audubon Magazine's blog, 2011 Holiday Gift List

"*Feathers* should appeal to a wide range of readers interested in science, natural history, or birds." —*Science Books & Films*

"At the risk of overusing the word, I have to say that the book itself is a kind of miracle too." —*Books and Beasts*

"A book well worth reading." —*10,000 Birds*

"Captivating." —*Science News*

"[*Feathers*] gracefully explores the evolutionary processes that led to the development of avian plumage." —*Seattle Magazine*

"[A] terrific natural history." —*Charleston Gazette*

"Wow! . . . [A] fun, wide-ranging romp." —*BirdWatching*

"[E]njoyable, wide-ranging, and well-researched. . . . Highly recommended for birders and science buffs."

—*Library Journal*, starred review

"Readers from science buffs to those interested in cultural history will find this a worthwhile afternoon's read." —*Publishers Weekly*

"[A] fascinating and eminently readable exploration." —*Booklist*

"A delightful ramble through the byways of evolution and the wonderful world of birds." —*Kirkus Reviews*

"[E]ngaging. . . . For all the intriguing science, what really livens up Hanson's passionate discussion of his 'natural miracle' are the stories he tells." —*Maclean's* (Canada)

"*Feathers* is a historic and truly interesting insight into how an everyday object can be marvelled at." —*Cage & Aviary Birds* (UK)

"Damn good book." —*BBC Wildlife* (UK)

"*Feathers* is simply a splendid book!"
—Robert Michael Pyle, author of
Wintergreen and *Mariposa Road*

"This is science written in clear and entertaining prose; a great read."
—Bernd Heinrich, Emeritus Professor
of Biology, University of Vermont;
author of *Winter World* and *Mind of the Raven*

"This is rich and engaging ornithology at its best."
—Frank B. Gill, author of *Ornithology*

"Absolutely fascinating history, and a terrific read."
—Garth Stein, author of *The Art of Racing in the Rain*

"A fascinating book . . . unusually well-written. Highly recommended."
—Peter Matthiessen, National Book Award-winning
author of *The Snow Leopard* and *Shadow Country*

FEATHERS

The Evolution of a Natural Miracle

THOR HANSON

BASIC BOOKS
A Member of the Perseus Books Group
New York

For Eliza

Hardcover first published in 2011 by Basic Books,
A Member of the Perseus Books Group
Paperback first published in 2012 by Basic Books

Designed by Trish Wilkinson
Set in 11 point Goudy Old Style

The Library of Congress has catalogued the hardcover as follows:

Hanson, Thor.
Feathers : the evolution of a natural miracle / Thor Hanson.
p. cm.
Includes bibliographical references and index.
ISBN 978-0-465-02013-3 (hardcover : alk. paper)—
ISBN 978-0-465-02346-2 (e-book) 1. Feathers. 2. Birds. I. Title.
QL697.4.H36 2011
598.147—dc22 2011003272

ISBN 978-0-465-02878-8 (paperback)

10 9 8 7 6 5 4 3 2 1

Contents

Flight

Fancy

Function

Author's Note

Throughout this book, birds are referred to using English names standardized by the International Ornithological Union. By this convention, individual species are capitalized (e.g., Winter Wren, Lesser Bird of Paradise), while groups of birds or generic references are not (e.g., the wrens, the birds of paradise). The IOC species list is maintained online and updated regularly by an international panel of ornithologists (see Gill and Donsker 2010). It eliminates the need to clutter up the text with endless parenthetical Latin binomials. I've similarly avoided in-text citations in favor of trailing phrase notes, which identify quotations and highlight important source material for each chapter. See the notes section, which begins on page 283. A complete list of references is included in the bibliography.

Acknowledgments

Writing a book of this kind relies on the generosity of others. From scientists and museum curators to fishing guides and fashion designers, people throughout the world of feathers have come to my aid time and again—assisting with research, answering questions, sharing data and specimens, and sitting down for long feather-filled conversations. Here, in no particular order, are some of the people and organizations that have helped me along the way:

Bob Pyle, Hilda Boshoff, CP Nels Museum, Rob Nixon, Sarah Abrevaya Stein, Jodi Favazzo, Marios Ignadiou, Greg Willson, Scott Hartman, Xing Xu, Wyoming Dinosaur Center, Alan Feduccia, Richard Prum, Carla Dove, Pete Menefee, Rainbow Feather Company, Marian Kaminitz, Leah Chalfen, Simon Thomsett, Laila Bahaa-el-din, Anziske Kayster, Graaff-Reinet Museum, the family of Russel W. Thornton, Peter Liotta, National Audubon Society, Kathy Ballard, Kim Bostwick, Patrick Kirby, John Sullivan, Tony Scruton, Bernd Heinrich, Edward Bormashenko, Yanchun (Daniel) Xu, Peter Harrison, Julian Vincent, Ken and Suzanne Franklin, Suzanne Stryk, Shirley Reuscher, Ken Dial, Ellen Thaler, Angela Linn, University of Alaska Museum of the

North, Smithsonian Feather Identification Lab (National Museum of Natural History), National Museum of the American Indian, Pacific Coast Feather Company, Jeffrey Long, Travis Stier, Glenn Tattersall, Brenda Boerger, David Houston, Maureen Goldsmith, University of Pennsylvania Museum of Archaeology and Anthropology, Elana Kitching, Linda Lorzechowski, Glenn Roe, ARTFL Encyclopédie Project, Tom Whiting, Whiting Farms, David Roberts, Cliff Frith, Jane Grayer, Ian Strange, Donald Jackson, Chris Strachan, Robert Petty, Impington Village College Model Aeroplane Club, Peter Stettenheim, William T. Cooper, Robert Petty, Theiunus Piersma, Max Platzer, Gwen Bisseker, Nickolay Hristov, Airling Gunderson, and Petra Quillfeldt.

Special thanks are due to Paul and Ann Hanson and Erin Braybrook for some very timely child care. I'm also deeply indebted to the entire staff of the San Juan Island Library, and particularly Heidi Lewis for her patience with my endless interlibrary loan requests.

Frank Gill kindly lent his expert eye to early drafts of the manuscript and provided valuable and encouraging feedback.

I'm grateful to my agent, Laura Blake Peterson, for shepherding a field biologist through the world of publishing and for connecting me with T. J. Kelleher at Basic Books. T. J.'s editorial savvy and enthusiasm for this project have been a boon. Whitney Casser, Cassie Nelson, Sandra Beris, Annette Wenda, and the rest of the team at Basic have also been a pleasure to work with.

Finally, none of this would have been possible without the unwavering support of my wife, Eliza Habegger. Along with our baby boy, Noah, and the rest of my family and friends, she patiently puts up with my quirks, queries, travels, and lengthy retreats to the Raccoon Shack.

Preface

O! I am Fortune's fool.
—William Shakespeare,
Romeo and Juliet (ca. 1595)

Vultures made me do it. That's my stock answer now, whenever people ask me about this book. It was vultures that first spurred my interest in feathers, years ago on a research project in Kenya. Watching the great birds hiss and squabble around a carcass, I thought of how perfectly their feathers (and lack thereof) were suited to the lifestyle. Bare heads and necks provided for cleaner feeding as well as heat regulation, stretched out long to keep cool during the day and tucked back into a plush, downy collar at night. Their dark body plumage resisted bacteria and absorbed the hot African sun, helping them stay warm in the chilly high altitudes where they soared, searching for the next kill.

The vultures started me thinking about feathers, and I've never stopped. I've seen flycatchers and nightjars burdened with breeding plumes three times their body length and watched penguins plunge beneath ice floes, comfortably watertight inside their satiny coats. I've huddled in a goose-down sleeping bag on subzero nights, while the tiny kinglets I studied kept perfectly warm nearby, fluffed up against the icy winter wind. I've traced feather shapes in the stone of dinosaur fossils and seen them in flying machines, fishing lures, Victorian hats, shuttlecocks, fletching, and ancient Peruvian artwork. As ornithologist Frank Gill observed in his classic textbook, *Ornithology*, "The details of feathers have fascinated biologists for centuries; it is an enormous topic." Perfect for a book, I'd often thought, but it would take another vulture to set me on the path.

Let me explain. As a field biologist, I'm never at a loss for things to study or topics to write about: everything in the natural world is fair game. If I'm not intrigued and excited every time I step outside, it just means I'm not paying attention. Some people find it excruciating to go for a hike with me and my constant distractions: bird nests, butterflies, lichens, ant hills, soil types, bug frass, rocks—you name it. At home, my wife, Eliza, puts up with dead voles and songbirds tucked into the freezer; plant specimens filling the fridge; boxes of unidentified bees, old bones, and owl heads; and a big tank full of interesting grubs. (Our baby, Noah, puts up with a lot, too, but he's never known anything different!) I'm a fundamentally curious person, and it's never hard finding subjects of interest; the challenge lies in narrowing them down.

In the world of scientific research, competition for funding quickly eliminates most possibilities. Science takes money, and you need a timely, sexy topic to pick up grants. It's not surprising that whales and tigers get more attention than liverworts, click beetles, or mold. Basic field biology can be a tough sell, and I usually frame my work in the context of larger themes: habitat fragmentation, species conservation, population genetics, or even the ecological impacts of warfare. When my schedule finally opened up to start a new book, however, I found the range of potential topics almost overwhelming. On the first morning, I sipped coffee and stared at an empty page before finally starting a vulture story I'd been meaning to jot down for years (you'll find it in Chapter 15). I hoped it would at least get the creative juices flowing, and it might come in handy if I ever wrote "the feather book."

I'm not the world's fastest scribe, but I had a few rough paragraphs by the time I broke for a midday run. We live on an island, five miles from town on a country lane that slopes gently downhill through dense woods and out between two farm fields. As I jogged along, thinking about vultures and feathers, my nose registered the growing rot-and-copper reek of a dead animal. I entered a stand of trees, and there, sure enough, lay the upended rib cage of a road-kill deer, splayed out beside the ditch. Overhead, a young Bald Eagle kept vigil on a fir branch, and above him, higher in the same tree, sat four Turkey Vultures. They hunched together in a dark row, their red heads lowered, silent and staring.

I slowed, and the vulture on the end suddenly started up, flapping awkwardly, each wing beat a whistling strain for lift in the cool autumn air. It tilted and angled through the branches, banking sharply to gain the unobstructed sky above the roadway. As it passed overhead, I saw something drop from its left wing and drift, spinning, then wafting, then spinning again, to the ground at my feet. It was a flight feather—long, dark, and beautifully curved, lying there on the pavement like an open parenthesis.

Now, I'm a scientist and a bit of a skeptic. I don't read horoscopes, visit clairvoyants, or spend a lot of time worrying about fate. I do, however, have several friends capable of staging elaborate practical jokes. My first reaction was to look for the hidden camera, or listen for the sound of muffled laughter coming from behind a hedgerow. But of course there was nothing, just my breathing, the quiet woods, and the retreating flight noise of the bird. It really did appear that after spending the morning writing about vultures and their feathers, I'd gone for a run and bumped into a bunch of vultures, and that one of them had practically dropped a feather on my head.

"You don't choose what to write—it chooses you." I first heard that maxim intoned with great significance during an undergraduate creative writing seminar. At the time, it made me glad I had a double major in ecology, where I could balance such notions with a dose of comfortably prosaic tables, graphs, and data sets. Now, the phrase seemed less cliché than command. Ancient Egyptians revered vultures as far-seeing symbols of empire, truth, and justice, never to be denied. Fortunately, these birds had given me a mandate I was glad to fulfill. Decision made: I would write the feather book.

With a nod to the trio still perched in their fir tree, I picked up the feather and carried it home. It's here with me now, the vulture's benediction, token of an exploration just beginning, and a fascination that will never end.

INTRODUCTION

A Natural Miracle

Lewis stoops to pick up a red-tinged feather lying on the path. He tells me that it belongs to a flicker, points out some of its features—rachis, vanes, calamus—then, giving it to me, says that I now hold in my hand a natural miracle.

—Leonard Nathan,
Diary of a Left-Handed Birdwatcher (1999)

I walked in the lead as the group turned onto a path by the field's edge, stepping softly in the dew-wet grass. Our shadows stretched westward in the morning sunlight, crazily adorned with binocular shapes, tripods, and the long limbs of spotting scopes. It was the first spring field trip for the local Audubon club. We had started with great blue herons and a pair of yellowlegs patrolling the tide flats and were slowly making our way inland toward a freshwater marsh, where I knew the wood ducks had recently returned from migration. Scattered milk-white clouds scudded through the blue overhead, and the sun felt warm against our faces, a strange but welcome sensation after the rain-soaked Pacific Northwest winter.

My eye caught a flutter of wings and a flash of russet near the fence line. I raised my binoculars, and the bird came into clear focus, standing alert in the short, green grass. "There's a —," I began, but my mind went blank. The group stopped, and I sensed everyone turning to look, lifting binoculars, and setting up scopes. It was an obvious bird, really, hardly worth mentioning to a group of pros like this. "By the fence there. It's a —." I reached for the name again, but got nothing. A mental dial tone.

"It's a *robin*," the man next to me said acidly, lowering his binoculars. The others turned away too, and there was a moment of awkward silence. I was leading an Audubon Society field trip, and I had just forgotten the name of the American Robin, possibly the most common backyard species on the continent. In the bird-watching community, this was a faux pas akin to an astronomer's forgetting the name of the Earth. Just then I heard someone say "Warbler!" and the group hurried off up the trail. With my credibility pretty much shot, it was a relief to stay behind and watch the robin.

The subtle rust and charcoal hues of her plumage told me it was a female, and her feathers shone fresh and porcelain smooth in the sunlight. She cocked her head, hopped, and then lunged forward to root at something in the soil. Tilting upright again, she suddenly launched skyward, turning sharply around a fence post and swooping up at an impossible angle to land on an alder branch. Perched there, the robin shook her tail and fluffed up her body feathers before letting everything settle back into place. Then she began to preen, turning and dipping her beak to lift and comb individual quills and vanes, like a fussy housekeeper arranging and rearranging the furniture.

American Robins, by John James Audubon.

I smiled, but who could begrudge her perfectionism? Those feathers impacted every aspect of her life. They protected her from the weather, warding off sun, wind, rain, and cold. They helped her find a mate, broadcasting her femininity to any male in the neighborhood. They kept out thorns, thwarted insects, and, above all, gave her the skies, allowing a flight so casually efficient that our greatest machines seem clumsy in comparison. Abruptly satisfied with her plumes, the robin dropped from the branch and set off over the field, wings parting the air in quick, certain strokes. I lowered my binoculars, far behind the Audubon group now, but glad to have been reminded of a natural miracle, feathers, as common around us as a robin preening and taking flight.

On any given day, up to four hundred billion individual birds may be found flying, soaring, swimming, hopping, or otherwise flitting about the earth. That's more than fifty birds for every human being, one thousand birds per dog, and at least a half-million birds for every living elephant. It's more than four times the number of McDonald's hamburgers that have ever been sold. Like the robin, each of those birds maintains an intricate coat of feathers—from roughly one thousand on a Ruby-throated Hummingbird to more than twenty-five thousand for a Tundra Swan. Lined up end on end, the feathers of the world would stretch past the moon and past the sun to some more distant celestial body. Their exact number is unknowable, but one thing is certain: from the standpoint of evolution, feathers are a runaway hit.

Animals with backbones, the vertebrates, come in four basic styles: smooth (amphibians), hairy (mammals), scaly (reptiles, fish), or feathered (birds). While the first three body coverings have their virtues, nothing competes with feathers for sheer diversity of form and function. They can be downy soft or stiff as battens, barbed, branched, fringed, fused, flattened, or simple unadorned quills. They range from bristles smaller than a pencil point to the thirty-five-foot breeding plumes of the Onagadori, an ornamental Japanese fowl. Feathers can conceal or attract. They can be vibrantly colored without using pigment. They can store water or repel it. They can snap, whistle, hum, vibrate, boom, and whine. They're a near-perfect airfoil and the lightest, most efficient insulation ever discovered.

Standing there, watching my robin, I was hardly the first biologist enthralled by a feather. Natural scientists from Aristotle to Ernst Mayr have marveled at the complexity of feather design and utility, analyzing everything from growth patterns to aerodynamics to the genes that code their proteins. Alfred Russel Wallace called feathers "the masterpiece of nature . . . the perfectest venture imaginable," and Charles Darwin devoted nearly four chapters to them in *Descent of Man*, his second great treatise on evolution. But the human fascination with feathers runs much deeper than science, touching art, folklore, commerce, romance, religion, and the rhythms of daily life. From tribal clans to modern technocracies, cultures across the globe have adopted feathers as symbols, tools, and ornaments in an array of uses as varied and surprising as anything in nature.

At Chauvet Cave in southern France, there is a Long-eared Owl carefully etched into the soft stone of the ceiling. Simple deft lines show the bird looking backward over its feathered shoulder in an unmistakably owlish posture. The image is one of thousands, a minor piece in the collection of petroglyphs and pictographs that make Chauvet, Lascaux, and other nearby caverns a treasure trove of prehistoric art. Haunting and evocative, their ancient animals, designs, and figures are crafted with such skill they moved Pablo Picasso to lament, "We have learned nothing in 12,000 years." In fact, the artwork at Chauvet dates back more than 30,000 years, making that small owl the world's oldest known depiction of a bird.

Although artifacts from the period include delicate bird-bone needles, flasks, beads, and pendants, individual feathers are rare in these early cave paintings. Archaeologists believe that ancient hunters used feathers, too, for ornamentation and as brushes for their ocher paints. By the late Stone Age, feathered headdresses and fletched arrows appeared regularly in rock and cave art from Europe to the American Southwest to the deserts of Namibia. Already, people had co-opted feathers for uses both practical (to make an arrow fly true) and deeply cultural (as prized adornments for ceremony and status). Their varied, often vibrant colors made feathers an obvious choice for decoration. Before modern pigments, what other medium offered everything from the beige and

A Long-eared Owl at Chauvet Cave, southern France.

umber of pheasants to the bright iridescence of sunbirds, motmots, and parrots? In time, feathers would spawn a global industry, clothe kings and courtesans alike, and define the height of fashion from Paris to New York. The use of feathers for fletching marked a similarly intuitive leap, from flight observed to flight intended. Indeed, their durability and aerodynamic structure would inspire engineers and inventors from da Vinci to the Wright brothers. The consistent appearance of feathers in myth and ritual, however, points to deeper mysteries.

When Emily Dickinson wrote, "Hope is the thing with feathers / that perches in the soul," she echoed an age-old sentiment linking feathers and bird flight with a sense of portent, longing, and the spirit. In ancient Rome, official fortune-tellers called augurs based their predictions on the behaviors of birds or on patterns seen in their feathers, bones, and viscera. These bird oracles held great sway, influencing major decisions in politics as well as private life, and even today we recall the past importance of augury when we inaugurate presidents or speak of an auspicious occasion. Syrians, Greeks, and Phoenicians divined omens from the cooing of doves, and mystics from many traditions have described the soul or the path to enlightenment in avian terms. To the Sufi poet Rumi, the human spirit was alternately a parrot, a nightingale, or a white falcon on a spiritual journey to God: "When I hear Thy drum . . . my feather and wing come back." In central Asia, the Dolgan people described the souls of children as tiny birds perched in the Tree of Life, while shamans from South America to Mongolia have described their trancelike states as "riding the wind." Near-death experiences invariably feature a disembodied phase, looking downward from a bird's-eye view, and both Jung and Freud considered flying dreams among the most powerful (though whether they symbolized transcendence or rowdy sex was a point of debate).

To earthbound humanity, the ability to fly is inherently otherworldly, revered for its sheer proximity to the heavens. And if flight is sacred, then birds, wings, and feathers are its most potent symbols, appearing again and again in a dizzying range of rituals, beliefs, and customs. Birds and bird-gods figure heavily in all mythologies, and flight is the jealously guarded privilege giving them access to both the spiritual and the earthly planes. In ancient Greece, Hermes relied on winged sandals to speed his passage to and from Mount Olympus, but when the mortal boy Icarus flew too high, his wax and feather wings fell to pieces. The Hindu messenger god, Garuda, emerged from an egg with the body of a man and the plumage of an eagle. Flight earned him the honor of transporting Vishnu and gave him eternal advantage over his devious serpent-spirit adversaries, the Naga. Revered by Buddhists as well as Hindus, his wildly feathered visage still adorns the official seals of Thailand, Indonesia, and Ulaanbaatar.

In some traditions, feathers are a symbol of spiritual purity and a prerequisite for an agreeable afterlife. Upon their death, ancient

Egyptians believed that the jackal-headed god Anubis would measure the worth of their heart, and the soul it contained, against the weight of a feather. Those found in balance entered the pleasant kingdom of Osiris. But when the scales tipped wrong, Anubis flung the offending heart into the waiting maw of Amemait, "the Devourer," a slavering hippo-lion-crocodile beast that crouched at his feet. In the Peruvian Amazon, the Waorani people also faced a feathery judgment at death, as described by ethnologist Wade Davis in his book *One River*: "Each Waorani has a body and two souls. . . . [T]he one lodged in the brain ascends to the sky where it meets a sacred boa at the base of the clouds. If and only if its nostrils have been pierced and decorated by the finest of feathers can the soul enter heaven. If turned away, it falls back to earth and is consumed by worms."

The connection between feathers and the sacred did not stop with shamanism or ancient mythologies but found firm footing in the great monotheistic faiths as well. Christianity, Islam, Judaism, and even Zoroastrianism all share a belief in angels, higher spiritual beings that serve as intermediaries on the path toward unity with God. Over the centuries, the depictions and descriptions of angels have been surprisingly consistent. They feature clearly recognizable human figures augmented by the addition of certain features. And what was added? Just what was given to the human form to symbolize an elevated, angelic state? More hair? Scales? A coating of sticky amphibian slime? No, ever since Vohu Manah first appeared to Zoroaster, Michael to Moses, and Gabriel to Muhammad, angels have come equipped with great feathered wings. And the feathers are diagnostic—these are not the leathery, bat-like appendages featured on demons or the devil.

Like Hermes before them, angels used their gift of flight to pass from heaven to earth and back again, often bearing divine tidings. For some, their wings and feathers formed an elaborate pedigree, a

Mosaic of a six-winged, elaborately plumed seraphim, from the twelfth-century Chapel of the Angels at Mont Sainte Odile, Alsace, France.

sign of status. The chubby little angel haunting a Renaissance mural might boast only two short, stubby wings, while 6, 36, or even 140 pairs appear in various depictions and descriptions of the archangels. At the highest sphere, a seraphim's feathers were said to resemble peacock plumes, adorned with hundreds of all-seeing eyes. Texts like the Old Testament's Psalm 91 even attribute feathers directly to the Almighty: "He shall cover thee with his feathers, and under his wings shalt thou trust: his truth shall be thy shield and buckler."

Truly, the human fascination with feathers is as rich as their natural history. Any thorough exploration must span the sacred and the secular, the practical and the fantastic, from science to myth, culture, and art. Feathers give us insights into evolution and animal behavior but also provide a unique perspective on the history of human belief and ingenuity. Several main themes quickly emerge, providing a framework for the chapters of this book. *Evolution* explores the contentious mystery of feather origins—where did they come from, and why? *Fluff* investigates the amazing insulating quality of feathers, from tiny birds in an ice storm to the down in a mountaineer's parka. *Flight* reveals how feathers opened up the skies, and *Fancy* tells the exotic story of allurement, from birds of paradise to showgirls on the Vegas Strip. A final section, *Function*, investigates how feathers continue to evolve, both in nature and in the myriad additional ways they've been adapted for human use. Throughout the book we meet the creatures and characters that bring the story of feathers to life, an eclectic cast of birds, dinosaurs, professors, milliners, inventors, explorers, and more.

As a writer, my job is to keep you turning the pages of this book, but as a biologist I encourage you to put it down once in a while. If you do, you'll soon find aspects of the story very much alive in the world around you. My wife remembers her grandmother saying, "You're never more than three feet from a spider." Even the best-kept home hosts scores of them, tucked into dark nooks and corners or hiding behind the walls. Well, you're never far from feathers, either. If they're not stuffing your pillows and parkas, they're covering every bird in every forest, field, backyard, suburb, and cityscape. You'll find feathers and their influence in fashion magazines, airplane wings, fishing lures, ballpoint pens, and fine art, but above all in the birds, those commonest of creatures so casually adorned with miracles. Go outside and watch them every chance you get. Look closely. You won't be disappointed.

Evolution

For the interesting point about a feather is really this, that it grew. It was not made in a moment, like a bullet poured red-hot into a mold: its little airy plumes, branched like a fern into tiny waving filaments, were developed by slow steps, piece after piece, and spikelet after spikelet. And what is true of this particular bit of down which I hold in my fingers, trembling like gossamer at every breath and every pulse, is also true of plumage as a whole in the history of animal evolution.

—Grant Allen, "Pleased with a Feather" (1879)

CHAPTER ONE

The Rosetta Stone

It's feathers I need, more feathers
for the life to come. And these iron teeth
I want away, and a smooth beak
to cut the air. And these claws
on my wings, what use are they
except to drag me down, do you imagine
I am ever going to crawl again?

—Edwin Morgan,
The Archaeopteryx's Song (1977)

Pick up a feather and run it between your fingers. It feels light and soft, yet sturdy, the hollow quill tapering upward to a graceful vane. Whether it slipped loose from a gull's wing or escaped a down pillow, the design is unmistakable. We know immediately that it came from a bird—nothing else is so uniquely avian. Birds fly, but so do bats and mosquitoes. Birds lay eggs, but so do fish, newts, and crocodiles. Gorillas make nests, cats and crickets chirp, and squid have beaks. Of all the conspicuous traits and behaviors that make a bird a bird, only feathers are theirs alone. So where did they come from? Fossils tell us the evolution of feathers took place deep in the Mesozoic era, closely intertwined with the origin of birds themselves. The question remains one of the most intriguing and contentious debates in science, and it begins with the story of a fossil, a dowry, and a famous battle of wits.

It all started with a cough. For a nineteenth-century quarryman, coughing was nothing new. He worked in a world of dust, the fine grit of limestone powdered by blasting, chisel work, and constant hammer blows. Some of the men tried to escape it, covering their faces with cloth masks, but everyone still coughed, joking darkly that miner's lung would get them in the end, if the black damp or a wrong-cut slab didn't do the job first. The dry summer months were the worst—in August he might not have thought twice about a little wheezing or a persistent hack. But this was springtime in Bavaria, when rain and late snowmelt kept the quarry walls slick and dampened the dust to mud underfoot. His cough was something else, and surely he feared tuberculosis, a disease known at the time as *consumption* or *the white plague*. In nineteenth-century Europe, TB killed more people than any other disease, and everyone had watched friends or relations succumb to its wasting effects. Whatever the quarryman thought, he eventually took the drastic, expensive step of seeing a doctor.

The year was 1861, and trained physicians were a rarity in the German countryside. They mostly served the upper classes: wealthy landowners, merchants, nobles, and high clergy. Few stonecutters could afford their attentions, but those who worked the quarries around the village of Solnhofen had an advantage. The limestone slabs they prized from the earth contained more than paving blocks or plates for lithography. A careful split along their fine-bedded planes sometimes revealed dark skeletal prints of fish, leaves, insects, or strange creatures that no one had ever seen. With a wave of interest in natural history and Charles Darwin's recently published theories sweeping Europe, such fossils had been transformed from mere curiosities into valuable commodities. Museums and private collectors competed hotly for choice specimens, so the quarry owners started claiming any finds for themselves, regarding them as an important new revenue stream. The workers, however, had long regarded fossils as a rare perk in a dangerous and underpaid job. They still smuggled them out whenever they could, tucked into a coat pocket or hidden in a lunch pail. Everyone knew that the old doctor in nearby Pappenheim was an avid collector, and while he might not buy your fossil outright, he would always take good specimens as payment for medical treatment and advice.

Our quarryman had a serious chest infection when he finally reached the Pappenheim clinic, but the doctor's notes do not tell us his prognosis, his treatment, or even his name. We know only his method of payment: a delicate, crow-size fossil that would change science forever. It was the first full specimen of *Archaeopteryx lithographica*, an ancient animal with the skeleton of a reptile and the feathers of a bird. Though its Latin name was lucid, even poetic, "ancient wing, written in stone," there was nothing simple about its reception. The combination of reptilian and avian features ignited a firestorm, fueling ongoing controversies about evolution, creationism, and the origins of birds and feathers. More than 150 years and a thousand research papers later, *Archaeopteryx* is arguably the most scrutinized specimen in history, a contentious linchpin of evolutionary thought known to many as biology's "Rosetta stone."

To Dr. Carl Häberlein, it was a windfall. In recent years, the fossil trade had become an increasingly important part of his livelihood. Though untrained in paleontology, he was a recognized expert in the preparation and identification of Solnhofen fossils. His personal collection numbered in the thousands, and scholars often visited Pappenheim to view his exquisite pterosaurs, fish, and winged insects. He'd sold specimens to the top museums in Europe and earned a reputation for shrewd, hard-nosed negotiation. In *Archaeopteryx* he knew he had a fossil that would attract interest, spark controversy, and, most important, fetch a high price on the open market.

The only surviving photograph of Häberlein shows a pale, square-faced man with dark eyes and a thin-lipped smile. He is seated, hands folded primly, wearing a black doctor's frock and staring directly into the camera. It's easy to imagine him driving a hard bargain. With maximum profit in mind, Häberlein decided not to sell the fossil by itself; anyone interested in *Archaeopteryx* would have to buy his entire collection. He created an aura of mystery by refusing to loan the fossil or allow it to be sketched. Instead, interested buyers were invited to his home for brief private viewings. News spread quickly to key players at the Bavarian State Museum, but after months of haggling, they failed to agree on a price.

When a description of the fossil reached London, however, Häberlein found a determined bidder in Sir Richard Owen, director of natural history at the British Royal Museum. Owen's board of trustees balked at the price, but he and a colleague defied them, conducting secret negotiations with the doctor that dragged on for half a year. For Häberlein, the stakes were high: he was seventy-four years old, a widower, with a daughter at home who required a substantial dowry to marry within her station. Family honor (and a comfortable retirement) required a lucrative sale. For Richard Owen, the stakes were even higher. He was the preeminent paleontologist of his day, an adviser to Queen Victoria, and the man who invented the very word *dinosaur*. He'd built his career, however, on the firm belief that species were created and altered only by the hand of God. If *Archaeopteryx* could be perceived as an intermediate step between reptiles and birds, it would be dangerous fodder for the Darwinists. They would call it evidence that birds and their most distinctive feature, feathers, *evolved* from the reptiles. Owen needed to be the first person to study the fossil, to describe it, and to refute any possibility that it was a "missing link."

With his reputation at stake, Owen blinked first. He agreed to Häberlein's price of seven hundred pounds, the equivalent of nearly sixty-five thousand pounds (one hundred thousand dollars) today. It was a princely sum that guaranteed Häberlein's daughter a fine wedding, and while the trustees of the British Museum were furious, the price tag later seemed a bargain for what one paleontologist has described as "the most valuable specimen of anything, anywhere."

When the *Archaeopteryx* arrived in London, double-boxed in stout wooden crates stuffed with straw, it landed in a world where scholarly debates were headline news. Darwin's *On the Origin of Species* was not yet two years old, and the provocative idea of evolution by natural selection reverberated from lecture halls to parlors and even public houses. Everyone had an opinion. Political cartoons depicted monkeys in cradles and gorillas dressed for the office, while public forums drew large, raucous crowds that often shouted down the keynote speakers.

Darwin's theories heralded a fundamental shift in scientific thought, but that was far from clear at the time. His supporters, led by the charismatic Thomas Huxley, faced stiff opposition from established scholars like Owen, as well as the majority of the public. The thought of an unguided natural process shaping the course of life on earth flew in the face of church teachings and two thousand years of Western science and philosophy. Still, the intellectual appeal and explanatory power of Darwin's ideas were steadily

Archaeopteryx arrived in London when debate over Charles Darwin's ideas was at its peak, and the great naturalist was often lampooned in the press.

gaining converts. Putting feathers on a lizard could tip the scale, and Owen knew it. He unpacked *Archaeopteryx* himself as soon as it reached the museum, spiriting the fossil away to his offices and rushing to publish a description. It's hard not to wonder if he suspected how that one feathered specimen would upend his career, taint his life's work, and put him forever on the wrong side of history.

The story of my *Archaeopteryx* is more prosaic. It arrived via Federal Express in a cardboard box, packed tightly in bubble wrap, newspaper, and Styrofoam peanuts. Unlike Victorian times, it doesn't take the influence and bankroll of the British Royal Museum to find a specimen—I bought mine on eBay. Of course, it wasn't real: only ten *Archaeopteryx* fossils have ever been uncovered. Mine was produced from a cast of one of the originals, a high-quality copy intended for display in museums, classrooms, and private collections. These replicas sell well, and I can understand why. Books and research papers may feature great photographs of *Archaeopteryx*, but you can't feel the texture of a photo; you can't tilt it in the sunlight and watch shadows play across the shape of an ancient bone. I'm a field biologist—I like to get my hands on things.

When visitors to the Louvre view da Vinci's *Mona Lisa* for the first time, they're often startled by its size. They expect the painting to be much bigger. The same can be said for *Archaeopteryx*. By reputation, it should be a condor, an eagle, or a feathered dragon. In reality, it's more like a magpie or one of those drab hedge-dwellers that birders lump together as "Little Brown Jobs." After all I'd read about *Archaeopteryx*, I knew the fossil would be small. What surprised me was its beauty.

The artist who finished my replica captured the golden hues of Solnhofen limestone perfectly, and the fossil stood out from the slab in a deep, rusty brown. Its neck was arched and wings splayed as if it had glided calmly to its final resting place in the Jurassic mud. I could see individual bones, claws, delicate teeth, and of course the feathers. They surrounded the tail and curved back from the wings like the brushstrokes of a Japanese calligrapher. Up close, each rachis and vane looked fully modern, indistinguishable from those adorning the Little Brown Jobs in my yard. The level of detail was exquisite and the combination of avian and reptilian features obvious. Even to an untrained eye, it was easy to see what all the fuss was about.

It may be simpler to get a look at *Archaeopteryx* in the twenty-first century, but one thing clearly hasn't changed in 150 years: the controversy that surrounds it. Two castings were for sale on eBay when I made my bids, and their item descriptions could have been written by Huxley and Owen themselves. The first seller noted, "It has long been accepted by most scientists that

The author's cast of *Archaeopteryx lithographica*, the "ancient wing written in stone."

Archaeopteryx was a transitional form between birds and reptiles." The second was adamant: "*Archaeopteryx* could NOT have evolved from dinosaurs. . . . [C]reation predictions in all areas of science have been, and are being confirmed; while the predictions of evolution scientists are consistently refuted just as they have been in the case of *Archaeopteryx*." As a biologist, my sympathies lay with the Darwinian seller, but in the end the creationist had a better price.

Owen and Huxley lacked eBay to carry out their arguments but found ample expression in the forums of the day. They had crossed swords many times, on everything from seashells to fish taxonomy, but it was the topic of evolution that would immortalize their feud. The two rarely debated face-to-face, but people knew how things stood when one of Owen's supporters famously asked Huxley whether it was his grandmother's or grandfather's side of the family that contained the gorillas. The scene was a packed auditorium at Oxford in 1860, less than a year before the discovery of *Archaeopteryx*. Huxley's retort was summarized by one observer as, "I would rather be the offspring of two apes than be a man and afraid to face the truth." In an age obsessed with pedigree and refinement, such a slanderous exchange sparked tumultuous shouts and jeers from the audience. One woman fainted and had to be carried from the room.

Against this backdrop of insult and intellectual tension, Owen sequestered himself with the *Archaeopteryx* fossil and worked feverishly on a monograph. Photographs from the period show him as hunched and owlish, with large, deep-set eyes and a serious mien. Huxley, by contrast, appears young and confident, his dark hair swept back neatly and his beard worn long in the Victorian style. Hindsight makes it tempting to view Huxley as the energetic advocate of new ideas and Owen as the stodgy defender of convention. But both men were active, accomplished academics with good reputations and popular support, and Huxley himself had initially been a vocal doubter of evolution. In fact, some historians believe his self-appointed role as "Darwin's bulldog" initially stemmed as much from a desire to oppose Owen as it did from any deep confidence in Darwin's theories.

Less than three months after *Archaeopteryx* arrived at the museum, Owen presented his findings at a meeting of London's Royal Society. *Archaeopteryx*, he concluded, was "simply the earliest known example of a fully formed bird." There was no missing link, no evidence of evolution, no transitional form. The fossil's reptilian features were incidental. It was an ancient bird, likely created by divine manipulation of the various long-tailed pterodactyls also known from the Solnhofen fossil beds. Case closed.

Though Owen's quickly prepared paper left many unanswered questions, it met with surprisingly little debate and went unchallenged for several years. In the end, however, his haste proved to be his undoing. Huxley, who had attended Owen's lecture, took his time in formulating a response. He worked first on an ambitious study of avian anatomy that revealed striking similarities between living birds and certain extinct dinosaurs. When he did turn back to *Archaeopteryx*, this new knowledge made his attack on Owen all the more devastating.

In a series of lectures and papers, Huxley soundly refuted Owen's treatment, carefully illustrating the skeletal characters that gave *Archaeopteryx* affinity to both birds *and* reptiles. But he went further, identifying a small dinosaur from Solnhofen, *Compsognathus longipes*, that resembled *Archaeopteryx* in everything but feathers. In one fell swoop Huxley illuminated not one but two defensible "missing links": *Archaeopteryx* with its clear similarities to both birds and reptiles and *Compsognathus* linking *Archaeoptery* to a specific reptile group, the dinosaurs. For the coup de grâce, he revealed that Owen had carelessly mistaken the position of the fossil in its slab. "It seems," he observed, "that Professor Owen cannot tell his left foot from his right." Praise for Huxley's analysis was exceeded only by Owen's humiliation. His reputation never recovered from the blow.

Of course, the controversy surrounding *Archaeopteryx* did not stop with Huxley's work. Periodic discoveries of additional specimens have kept the fossil in the limelight, and it's been making (and ruining) scientific careers for a century and a half. What's remarkable, however, is that in spite of all the studies and theorizing, the crux of Huxley's original 1868 thesis persists. Revived and expanded by Yale University's John Ostrom and others in the 1970s, the growing consensus view holds that *Archaeopteryx* and the birds evolved from dinosaurs, specifically from the meat-eating group called theropods, of which *Compsognathus* was a member.

With Huxley's prescience in mind, I read and reread his work, peering at my *Archaeopteryx* to pick out the details. I wanted to understand exactly what he saw, just how he intuited the link between birds and dinosaurs. My replica clearly showed the feathers, the reptilian tail, and even the "disunited metacarpals" in the hands. But his descriptions of the "widely arched form of the acetabular margin" or the "posterior surface of the external condyle of the femur" forced me to admit that either my replica lacked that kind of detail or, just as likely, I lacked the training to recognize it. To truly understand *Archaeopteryx* and the origin of feathers I needed two things: an actual specimen and a good paleontologist.

Until very recently, seeing an actual *Archaeopteryx* specimen close-up meant booking a trip to Europe and acquiring hard-to-get research permits. Häberlein's original specimen still resides at the Natural History Museum in London. Others can be found in major paleontological collections in Berlin, Munich, or Holland. But now there's another option: Thermopolis, Wyoming, population 3,172. There, by a strange turn of events that some have called paleontological piracy, the tiny Wyoming Dinosaur Center presides over one of the most beautiful and complete *Archaeopteryx* specimens ever discovered.

"People say we're in the middle of nowhere, and they have a point," admitted Greg Willson, the center's director of excavations. I suppressed an urge to correct him. I'd just driven across eight hundred miles of sagebrush and tumbleweeds, and I knew for a fact that I'd passed nowhere *long* before reaching Thermopolis. "But the bottom line is that this *Archaeopteryx* is more accessible than it's ever been, and if it wasn't here, it would probably be locked up in a private collection somewhere."

In his mid-thirties, with a broad smile and a thoughtful way of speaking, Greg grew up just down the road from Thermopolis and graduated from Hot Springs County High School. He moved away for college and was pursuing a Ph.D. in anthropology when this job at the Dinosaur Center opened up. "I jumped on it and moved back," he said. "My training was in human paleontology, so I've had some adjusting to do, but I love the work." After a pause he added, "Frankly, there aren't that many opportunities like this in Wyoming. If you want a job in science, you usually end up waiting around for someone at the university to die!"

We were eating grilled sandwiches at Pumpernick's Family Restaurant downtown, continuing our conversation after a morning spent poring over fossils. The story of the Thermopolis specimen sounded quite familiar. There were secret negotiations, an elderly seller, a cash-strapped German museum, and a beautiful specimen of mysterious origin. One hundred and fifty years after Häberlein, it was sale to the highest bidder all over again.

"No one even knew about this fossil," Greg explained. "It appeared when the widow of a Swiss collector offered it for sale to the Senckenberg Museum in Frankfurt. They couldn't come up with the money, and she wouldn't lower her price. If Burkhard hadn't stepped in, it would have been gone. She would have auctioned it on the private market."

He was referring to Burkhard Pohl, the eccentric Austrian founder of the Wyoming Dinosaur Center. Part scientist, part fossil impresario, Pohl came from a wealthy family (hair care, cosmetics) and trained as a veterinarian before transforming his passion for paleontology into a full-time vocation. Using his moneyed, fossil-friendly connections ("He moves in those circles," Greg told me), Pohl secured an anonymous buyer for the *Archaeopteryx*. The price remains a secret, but an inferior specimen sold for $1.5 million in 1999, and Pohl's client probably paid far more. A stipulation of the purchase entrusted the fossil to the center for curation, and the Thermopolis specimen was born.

Initially, the move caused a small outcry within the scientific community (or "ruffled a few feathers," as a reporter from *Science* magazine couldn't help but observe at the time). As a rule, paleontologists mistrust the commercial fossil trade, and many refuse to study anything in a private collection. Some critics felt that Pohl had hijacked the *Archaeopteryx* and that it belonged in a large public institution. Others simply worried that the Wyoming Dinosaur Center couldn't provide proper care and security, or that its remoteness would hamper access and study opportunities.

"Most of the commotion has died down now," Greg said. "Yes, we're a private museum, but we've made the specimen readily available. Dozens of people have come to examine it. Papers are being written. The information is getting out."

That information makes an even stronger case for Thomas Huxley's hypothesis, further linking *Archaeopteryx* with *Compsognathus* and the other theropods. The positioning and quality of the Thermopolis specimen show several key features that can't be seen in other specimens, and Greg walked me through them that morning.

"Our specimen has the best-preserved skull and feet discovered yet," he explained, pointing out fine details with the tip of a pen. We were looking at an immaculate first-generation casting—the real thing spends most of its time in a climate-controlled chamber behind glass. Like the replica I'd pored over at home, the Thermopolis *Archaeopteryx* is splayed out on the slab like a snow angel, bones and feathers neatly etched in bas-relief. But here the top of the skull was clearly visible, as if the animal had died glancing downward. Greg explained how the palatine bone, a tiny dimpled nub visible through a hole near the eye socket, had four prongs as in theropod dinosaurs, not three prongs as in modern birds. He showed me how the second toe was hyperextendable, like the killing claw made famous by *Velociraptor*, the ferocious theropod predator in *Jurassic Park*. And the hallux, the reversed toe that allows perching birds to grip a branch, here was only partially transposed and probably incapable of grasping.

We were upstairs at the dinosaur center, seated around a wide table filled with fossils and casts. Next to the *Archaeopteryx* sat Huxley's *Compsognathus*, and the two did look remarkably similar—the feet, the long tail, the toothy jaw. Of the ten known *Archaeopteryx* specimens, at least two were initially misidentified as *Compsognathus* until closer inspection revealed feather impressions or telling skeletal details. The table contained other treasures as well: Eocene birds, one hundred million years younger than *Archaeopteryx* and unmistakably avian, with toothless beaks, shortened tails, hinged ribs, and fully reversed halluces. Greg pointed out the reduction of digits, the fusion of wrist bones, and subtleties that built strong affinities from theropod to *Archaeopteryx* to bird.

"Now let's head down and see the big guys," he said. We descended a metal staircase "behind the scenes" to the first floor. I could hear the whir of machinery as we passed the prep room, where four technicians work year-round under fluorescent lights, chipping, cleaning, and cataloging specimens. When we emerged into the main hall, my eyes took a moment to adjust. Then I saw dinosaurs rising up in the dimness, some of them two stories tall: *Stegosaurus*, *Triceratops*, *Apatosaurus*, and many I didn't recognize, filling the hangarlike room. They were fully articulated skeletons, as fine as in any museum, but displayed with a no-frills style that matched the building's boxy metal exterior. Even Pohl had admitted the place "wasn't very pretty," but in a way it helped keep the focus where it should be: on the fossils. Numbering in the thousands, the assemblage started with donations from Pohl's private collection. But it continues to grow as other specimens emerge from the Jurassic mudstones right on the edge of town, where the center operates several dig sites on Pohl's seventy-five-hundred-acre ranch.

"We're the only museum in the world where you can actually help find the fossils," Greg told me, referring to their popular "dig for a day" program. It gives visitors a hands-on paleontological experience excavating some of the very fossils the museum puts on display.

"You'll want to see *Allosaurus*," he went on, pointing to a huge skull with curved, bladelike teeth. "That's the only theropod we've found on the ranch." But there were other meat eaters in the collection—a *Velociraptor* shown in a full sprint and even *Tyrannosaurus rex*, displayed in a lunging pose with its massive head thrust out and jaws open in midsnap.

"Look at its hallux," Greg said, pointing to a spike that jutted backward from the foot, just up from its three giant toes. "Look familiar? That's probably just how the *Archaeopteryx* walked, with the hallux held up off the ground like a dew claw." He mimicked the motion with his hands, three fingers forward and thumb turned partly back. He showed me other bird-theropod features as we continued the tour—*Velociraptor*'s fused wrist bones, *Oviraptor*'s furcula, like the fossils we'd seen upstairs grown large and broken free from their slabs.

The *Archaeopteryx* display sat in the corner, under the sweeping tail of a 109-foot plant eater called, appropriately enough, *Supersaurus*. It's hard to make a magpie look impressive in a room full of dragons, but the center had sprung for dramatic lighting and lengthy interpretive signs. A "family tree" showed the proposed relationship of *Archaeopteryx* to the dinosaurs, emerging from the theropod line under the title "First Bird." I had to laugh. Though Huxley may have won their long feud, Owen could claim one small victory: he called the fossil a bird, and so it remains.

It all made for a convincing story, but I couldn't shake the feeling that something was missing. After lunch I said my goodbyes to Greg and then returned to the center to take one last look. The room was cool and dark as I toured the displays alone, the flash from my camera sending wild skeletal shadows up the walls. There were no crowds to shuffle me along, so I lingered for a while at the *Archaeopteryx*, as if sheer proximity to the famed fossil might lend me a bit more understanding.

I was reminded of seeing King Tut's golden sarcophagus—not on one of its high-security, red-carpet American tours, but on a trip to Egypt years ago. There, at the National Museum in Cairo, all of Tutankhamen's treasures were kept in simple glass-topped cases overseen by a bored-looking guard with an AK-47. Visitors were free to stay as long as they liked, gazing, taking photos, or just contemplating the artifacts' enduring mysteries. Like Tut, *Archaeopteryx* left a lot of questions unresolved.

That's when it hit me: what about the feathers? There they were, clearly visible, spread out from each fossilized wing in neat, overlapping rows. Yet the story of *Archaeopteryx* was a tale told in bone: palatine, hallux, phalange, fenestra. For all my study of the "First Bird," I'd learned almost nothing about the feathers themselves. How did they evolve? From what? When? What were they used for? The fundamental questions remained unanswered.

CHAPTER TWO

Heat Shields, Gliders, and Insect Scoops

The sight of a feather in a peacock's tail, whenever I gaze at it, makes me sick!

—Charles Darwin ponders feather evolution, 1860, from a letter to Asa Gray

Rock crushes scissors. It's axiomatic, a truth established by countless school-yard matches of "Rock-Paper-Scissors." But what if the game were called "Rock-Paper-Feathers"? Well, it turns out that rock destroys feathers, too. Fossils of any kind are rare—the vast majority of creatures and plants die and rot away far from the accumulations of silt, ash, or other sediments that might preserve them—but feathers are exceedingly scarce. Like skin, hair, or soft body tissues, they face a double challenge on the road to fossilization. They decompose faster than bone or shell, and they're easily damaged by the heat and pressure required to form even the softest mudstone or shale. It's no coincidence that the Wyoming Dinosaur Center and other paleontology museums resemble giant bone yards; the hard bits are the ones most likely to survive.

In fact, the conditions necessary for feather fossilization are so unusual that for well over a century, *Archaeopteryx* stood virtually alone. Jurassic dinosaurs, pterosaurs, fish, insects, plants, and even early mammals continued to appear from sites around the world, but only Solnhofen produced any additional feathered fossils of similar age. And those were just variations on a theme: a handful of new *Archaeopteryx* specimens. In spite of great interest and thousands of research papers examining bird and feather evolution, the physical evidence remained incredibly thin.

Into this void, generations of paleontologists, ornithologists, biologists, and even chemists and physicists have launched their theories, thoughts, and conjectures. Along the way, the evolution of feathers has often become lost in arguments over the origin of birds and avian flight. The beautifully preserved feathers on an *Archaeopteryx* wing make conflating these stories almost irresistible, but it's important to keep the questions separate. How, why, and when feathers evolved may have less to do with birds than people usually think.

Traditionally, theories about the rise of feathers focused on the question of *why* they evolved, proposing some particular use as the original driver of their evolution. These are called functional theories, and chief among them was the idea that feathers evolved for flight, that their obvious aerodynamic traits could have developed only in the context of airborne creatures. Though evidence increasingly points to a more nuanced story, a vocal group still adheres to this view. Others have suggested that comparatively simple plumes must have predated the flight feathers seen in *Archaeopteryx*. These may not have been good airfoils at all but probably provided insulation or bore colors used for display and courtship. Some theories argue for the waterproofing or protective function of feathers, and one imaginative hypothesis pictures dinosaurs using feathered wings as heat shields to shade their eggs and hatchlings from the punishing Jurassic sun.

Even the best minds have succumbed at times to fancy. John Ostrom, who revived Huxley's theories and whose "hot-blooded" dinosaur ideas transformed the field, also proposed the notion of feathers as an "insect scoop." In this scenario, *Archaeopteryx* and other protobirds raced along, beating their feathered wing tips against the ground to scare up insects, or thrusting them in front of their open mouths like a great net, scooping bugs straight from the air. Though it made for a terrific illustration, even its author eventually admitted that such awkward lurching about was less likely to net a meal than a stumble face-forward into the mud.

Peppered with phrases like *trash*, *garbage*, and *quackery*, the debate on feather evolution is pretty saucy stuff for scientific literature. It can read like a tabloid drama, with each faction holding up some minutely measured detail of *Archaeopteryx* as its primary

John Ostrom's "insect scoop" theory featured ancestral birds using their feathers to trap flying insects.

evidence. Functional theories invite such wars of words because they share the same basic drawback: in the absence of a complete fossil record, or a time machine, they're almost impossible to test.

Feathers possess a range of highly adapted physical qualities, and no scientist doubts that natural selection has shaped and reshaped their design. The presence of modern-looking flight feathers on *Archaeopteryx* tells us that at least some of their varied functions were already fine-tuned by the late Jurassic. But the lack of earlier specimens leaves a gap into which functional logic stumbles. How can the first use of the first feathers be parsed out from the complex, fully developed traits of a later fossil?

Imagine you're an archaeologist from far in the future who finds a cell phone, just one, while excavating an early-twenty-first-century Earth site. Let's say it's an iPhone, a compact, glass-fronted device that can send e-mail, surf the Internet, take photographs, store and play music, identify pop songs, recommend a good steak house, and swiftly carry out a hundred thousand other applications. Without additional evidence, how could you deduce the history and original purpose of the device? Would you ever guess that its complex, multifaceted interface developed from a bulky box and receiver handset with only one function, sending voice messages over a copper wire? Even if that was your theory, you'd have nothing to support it unless you stumbled across a telephone from the twentieth century (or from the office of a certain Luddite field biologist).

Interpreting feather evolution from *Archaeopteryx* is much the same. The best specimens boast clearly identifiable wing and tail feathers and the suggestion of contour feathers on the legs. There is no reason to believe that down wasn't also present or that the feathers weren't as bright as those of a parrot or as cryptic as those of a quail. In fact, *Archaeopteryx* tells us little more about the origin of feathers than does an eagle, a penguin, or the sparrows futzing about in your yard. Just like their ancient ancestor, they too boast an array of feather types adapted to a range of functions. All *Archaeopteryx* really tells us is that modern feathers have been around for a long, long time. To know *why* they evolved may not be possible until we answer the question of *how*.

Clues to the how of feather evolution lie in their development, unraveling the way they grow into such an incredible diversity of forms. Hair and scales pale by comparison—the structural complexity of feathers surpasses any natural integument in the history of life. In size alone they can vary by several orders of magnitude *on the same bird*. The male Paradise Flycatcher, to pick one example, sports facial bristles that measure less than a millimeter and tail plumes that stretch more than two hundred times as long. When an Indian Peacock raises his tail, its glittering, iridescent coverts form a display more than fifteen hundred times longer than his shortest feather. If human hair were similarly diverse, a person might combine a neatly trimmed Van Dyke beard with a teased hairdo taller than the Statue of Liberty.

To truly get my head around this diversity of form and function, I decided I needed to dismantle a bird. I needed to see how the feathers grew, how many types there were, and just what differentiated one from another. This is the kind of idea that makes our chickens nervous.

My office occupies the southern half of what I call the Raccoon Shack, an old orchard shed we spruced up and named for its former inhabitants. The raccoons used to live underneath it, emerging on summer evenings to stage leisurely raids on the plum trees, apples, or anything else in season. These days it might be more appropriate to name the shack after our four laying hens. They can usually be found pecking and scratching right outside the door, occasionally hopping up on the porch to peer in at me, hunched over my books, computer, or microscope. Three of the hens were getting on in years, and I had my eye on Trouser, a surly old Wyandotte who'd never been much of a layer and spent most of her time plotting attacks on our lone Rhode Island Red.

Honestly, I wasn't sure I had the heart for premeditated chicken killing, and lucky for Trouser I didn't have to find out. The next time we defrosted our chest freezer I discovered, nestled between the halibut and old soup, a salsa bin labeled "Thor's Kinglet." Sure enough, it held a beautiful Golden-crowned Kinglet, a species we'll talk about in detail in Chapter 5. But right next to the kinglet sat a bonus, a tiny Winter Wren I'd forgotten about completely. Jackpot.

With its head cocked, eyes open, and stumpy tail turned up, the bird looked ready to leap from my hand. Of course, it was frozen solid, a roadkill I'd picked up near our house the previous winter. Like other wrens, this little brown skulker lived in dense underbrush and was probably hit trying to dart across the road from one thicket to the next. The old postal scale in my office told me it weighed less than a half ounce, and that in 1958 I could have mailed it anywhere in the continental United States for only four cents, first class. But Winter Wrens were far more accomplished travelers, having long ago crossed the Bering Sea to explore Siberia and beyond. The only wren species found outside the Americas, their range now extends throughout temperate Asia and Europe, even reaching the mountain forests of North Africa.

In terms of plucking, the wren offered several obvious advantages over Trouser. Measuring less than four inches from stem to stern, it was one of the smallest species in North America and seemed a manageable specimen for a novice plucker to tackle. Also, it was already dead and wouldn't involve testing my resolve for the ax-and-stump routine. Finally, plucking the wren would demonstrate once and for all that I do occasionally make use of the nasty dead things I'm constantly squirreling away around our house.

I thawed my wren on a stormy November morning and retired to the Raccoon Shack with the only reference on plucking that I

A Winter Wren.

could think of, *The Joy of Cooking*. Though Irma A. Rombauer didn't offer any recipes specific to wrens, she did have a section devoted to wild fowl where she strongly recommended dry plucking and noted, "It is much easier to draw and pluck a bird that is thoroughly chilled." The wren was still good and cold, so with a pair of tweezers and some needle-nose pliers, I was ready to go.

The world record for plucking a bird belongs to Vincent Pilkington of Cootehill, County Cavan, in Ireland. Before his retirement, Mr. Pilkington could defeather a whole turkey in one minute and thirty seconds, and he once plucked 244 of them in a single day. Two hours into my wren project, it was clear that I posed little threat to Mr. Pilkington's records. Feathers and bits of feather fuzz covered my desk, and the carcass *still* had quills and pinfeathers poking out from odd places, like some kind of morbid pincushion.

Of course, Pilkington hadn't been trying to count the feathers (208 per wing; 12 on the tail; more than 375 on the belly, breast, and back; and more than 400 on the neck, head, and face). He also hadn't bothered sorting them by type into neat little piles. Nor had he wasted additional time swearing and resorting those piles every time a stray elbow, draft, or sneeze scattered everything willy-nilly.

As for the bird, I'm glad I hadn't planned on serving it. Not only had I failed to achieve the tidy dressed look pictured in *Joy of Cooking*, but the bedraggled, skinny remnant wouldn't have made much of a meal. In fact, it would have left room on a toast point. If you ever find yourself plucking a songbird, you'll be amazed at how little is left when you're done. It speaks to the importance of feathers that in spite of their lightness, they exceed the dry weight of a bird's skeleton by two to one in most species. Naked, the wren looked impossibly small and utterly unlike the lively sprites that scolded me from the shrubs around our house. I couldn't help feeling I'd done the bird a great indignity and said a word of thanks before burying it under an apple tree.

The feathers, on the other hand, looked like works of art. I took one from the "left wing" pile and held it up to the light. From a tiny quill it broadened into a tapering vane, its leading edge brindled in chestnut and chocolate stripes. Backlit against a window, each individual barb in the plume stood out, divided again into tiny barbules that interwove to make a seamless, curving whole. It was asymmetrical, a primary flight feather with its rachis offset forward and widest point slightly toward the tip. Other piles held distinctly different varieties: contour feathers from the head and breast, tiny bristles from the face, and a large pile of dark semiplumes and down from the belly, so soft their touch was like the faintest brush of air, less a sensation than an anticipation.

In spite of such variety, every feather shares the same basic structure, rising from a hollow quill to branch from a central shaft, with the degree and pattern of divisions resulting in myriad feather types. Until recently, learning the details of feather structure and naming different feathers was a ritual reserved for undergraduate courses in ornithology. Flight feathers on the wing are remiges; tail feathers are rectrices. The central shaft in the vane is called the rachis. The base of the quill is the calamus. There are filoplumes, powderdown,

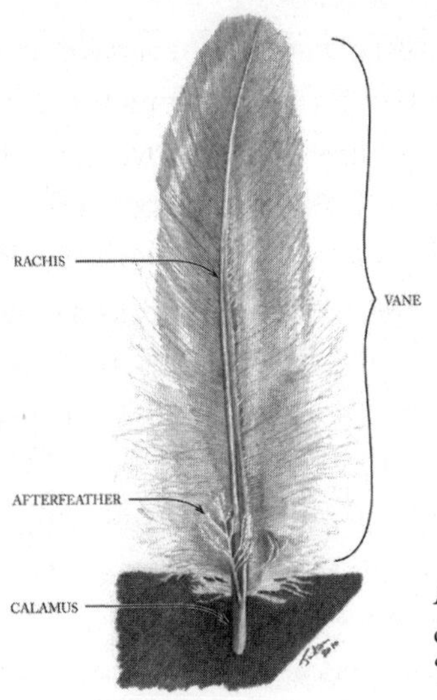

A typical contour feather. For examples of other feather types, turn to Appendix A, "An Illustrated Guide to Feathers."

wing coverts, tail coverts, and so on. Most students memorize these terms and forget them soon after the first exam. But a theory put forth by Dr. Richard Prum shows that learning how a feather *grows* may also answer the question of how feathers evolved.

"The realization actually came to me at the chalkboard," he explained. Energetic, with reddish hair and an informal manner that still conveys intensity, Prum likes to talk about feathers. We spoke several times on the phone and met once at his office, where piles of papers and opened books cover every surface like a hundred thoughts in process. He now holds the Coe Professorship of Ornithology at Yale, but he was teaching at the University of Kansas when the idea struck him. "I was giving my class the standard scales-to-feathers theory, and by the end of the lecture I realized it didn't make any sense. Feathers *couldn't* have evolved that way!"

Until Prum's epiphany, common wisdom held that feathers evolved directly from elongated scales that had frayed and divided in response to selection for flight (or whichever functional theory was in vogue at the time). The emphasis always lay on what feathers were used for, with scant attention paid to the mechanics of their transformation. What Prum recognized now seems obvious: there is a fundamental structural difference between scales and feathers and how they grow. Scales form like plates, flat ridges protruding outward as extensions of the epidermis. Feathers, on the other hand, are inherently tubular. It's like the contrast between a napkin and a straw. Fold the napkin and you have a scale, with the outer surface—the epidermis—covering both top and bottom. To flatten a straw you could of course smash it, but that's not how a feather grows. Feathers flatten by *opening up*. The outer surface becomes the top, and the inside is revealed to become the bottom. So while a mature feather and a scale may both appear flat, their surfaces simply don't correspond.

"I had just prepped a lecture on feather development," he told me, describing a late night in Kansas spent poring over old books. "I'd finally gotten it, how helical feather growth happens. It's an amazing and distinct biological process." When Prum talks, his boyish enthusiasm belies the distinguished professorship and his long list of academic honors. He explains things eagerly, as if willing you to understand everything he knows so that together you can take the ideas further. "From that moment, feather evolution became a big part of my classes. We did all of this jamming on a chalkboard."

Over time, Prum's jamming turned into the most lucid theory of feather origins to date. The theory is *developmental*, focusing on how feathers grow and not worrying about what they're used for. It hinges on one of the essential nubs of evolution: novelty. Evolution relies on the introduction of new traits to give natural selection and other processes something to act upon. No novelty, no change. The developmental theory identifies five unique attributes that had to be "invented" before the growth process could produce modern vaned feathers. This sequence of changes occurred within the feather follicles, those distinctive raised dimples of skin where feather growth takes place. Each novelty led to increased structural complexity, from an unbranched quill (Stage I) to simple branched filaments (Stage II), to the organization of filaments around a rachis (Stage III), to the development of interlocking barbules and pennaceous vanes (Stage IV), and finally to asymmetrical flight feathers (Stage V). Prum argues that each step required a new innovation in the follicle: barb growth led to a divided, downy feather; helical growth led to the rachis; paired barbs led to barbules; and so on. The growth of a fully modern, vaned feather would be impossible without the previous evolutionary steps.

The logic of Prum's theory is immediately appealing. It proceeds from the simple tubular quill, a universal feather feature, to increasingly complex forms. It remains neutral on function, concentrating instead on the novelties required for a feather to grow the way that it does. The theory's real strength, however, lies in its testability. Without hypothesis testing, even the most logical scientific ideas remain little more than conjecture. Support for functional theories has relied too often on traits or behaviors that leave no discernible fossil trace, but Prum makes specific testable predictions. If he's right, then every feather type produced by the five developmental stages should appear somewhere in the fossil record. As he puts it, the earliest feathers "need not precisely resemble any modern feathers, but should be plausibly grown by a conceivable feather follicle." Even more intriguing, and more controversial, is his prediction that finding these fossil feathers in the theropod dinosaurs will confirm the ancestry of modern birds.

Though Prum's research makes bold statements and some criticize his interpretation of feather growth, his ideas found quick acceptance, and their inclusion in every major ornithology textbook ensured that, in fewer than ten years, the developmental theory went from one chalkboard in Kansas to classrooms around the

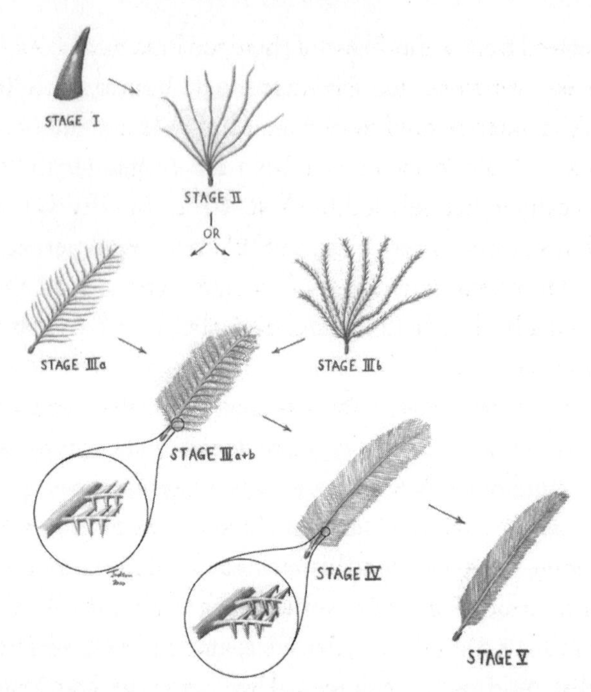

Developmental Model of Feather Evolution. The developmental theory proposes a series of cumulative evolutionary steps leading to modern feathers: an unbranched quill (Stage I), simple filaments (Stage II), filaments centered on a rachis (Stage III), interlocking barbules and pennaceous vanes (Stage IV), and asymmetrical flight feathers (Stage V).

world. This being an evolutionary question, however, particularly one involving feathers and birds, there remain a few doubters, skeptics, and outright naysayers.

"I'm not too keen on neutral evolution," Alan Feduccia told me over the phone. A longtime professor at the University of North Carolina, he spoke with an affable Southern drawl that seemed at odds with the sometimes strident tone in his articles on the topic. "Rick Prum is a bright guy and his theory looks good on paper, but does it really answer the question? I don't think so."

Nearly forty years ago, Feduccia found himself in the role of doubting Thomas, countering his friend John Ostrom's idea that birds evolved from warm-blooded theropod dinosaurs. It's a part he has played ever since, though increasingly from the sidelines, as Ostrom's views took hold to become what Feduccia calls "the new orthodoxy." A diminished but stalwart group agrees with him, informally calling themselves the BAND (Birds Are Not Dinosaurs). Though his positions now seem iconoclastic, Alan is anything but a crank. His scientific credentials include scores of peer-reviewed papers and a book, *The Origin and Evolution of Birds*, that is widely considered a classic.

"Maybe I'm too much of an old-school Darwinian," he went on, "but I don't see feathers evolving outside the clear contexts of adaptation and natural selection." Feduccia argues that novel features must be adaptive—if they don't confer some discernible benefit to the organism, then they're unlikely to persist. With that in mind he questions the usefulness of Prum's downlike Stage II feathers. As any camper in a rainstorm knows, down loses most of its insulative value when wet. Adult birds keep their down dry under layers of water-resistant contour feathers, and downy chicks must huddle beneath their parents to survive. Young ostriches caught out in the rain often die of exposure, even in the African heat. In Prum's model, however, contour feathers evolved *after* downy plumes, calling into question the notion that the first feathers were used for insulation.

"Feathers are unbelievable," Feduccia said, and his voice took on a tone of wonder I would hear again and again in my research—from scientists, hatmakers, engineers, fashion designers, even fly-fishermen. "They have all of these incredible aerodynamic features—lightweight, with graded flexibility; they're perfect airfoils; they can work together in slotted wings with high lift at low speeds. I just don't see how they could have evolved outside of an aerodynamic context." He adheres to the view that flight feathers evolved first, from scales, and that body feathers, down, quills, and all the other feather types came later.

"That's backward thinking," Prum dismisses, reiterating the intractable structural differences between flat scales and tubular feathers. Flight aside, early feather stages fit with any number of functional theories, from display to thermal regulation to touch, but choosing among them is highly speculative. "Concluding that feathers evolved for flight is like maintaining that digits evolved for playing the piano."

Until recently, Prum and Feduccia would have had little but their opinions to argue over. Feathered fossils remained limited to *Archaeopteryx* and a few late Cretaceous birds too young and fully avian to provide many clues.

"I might not have published when I did if Zhonghe Zhou hadn't taken my class," Prum recalled. After giving one of his feather evolution lectures in 1997, an exchange student had walked to the front of the room and told him how important his ideas were, practically pleading with him to write them up. The student was a paleontologist taking ornithology on the recommendation of his thesis adviser. What he knew that Prum didn't was that his colleagues back home were busy unearthing an unprecedented trove of feathered fossils.

After a century and a half with *Archaeopteryx* as the lonely linchpin in the feather debate, quills, down, and clearly vaned plumes began appearing on literally dozens of new specimens. The fossils were a revelation, as if a menagerie of perfectly plumed creatures had purposely plunged themselves into an ancient lake bed to die, beautifully preserved in the muck. They shook the scientific world. They allowed Prum's predictions and other ideas to finally be tested. And they also left anyone with an interest in the evolution of feathers or birds asking one another the same question: "How's your Chinese?"

CHAPTER THREE

The Yixian Formation

Usually the discoveries come piece-meal, one fragment at a time. . . . So paleontologists around the world were unprepared for the onslaught of fossils that came, and continue to come, from the fossil deposits of Liaoning Province in Northeastern China.

—Mark Norell,
Unearthing the Dragon (2005)

Xing Xu wanted to be a physicist. He idolized Niels Bohr and Albert Einstein, reading and rereading their theories in his high school textbooks and dreaming of following in their footsteps. Growing up in remote Xinjiang Province, not far from the Kazakhstan border, his educational opportunities were limited. "It was very backward there," he recalls. "To become a scientist was really just a fantasy." But he studied hard, and his test scores gained him entrance to Peking University, a prestigious school located thirty-six hundred kilometers to the east, in the heart of Beijing. He moved to the big city on a full government scholarship, with high hopes and the proud support of his family.

When he arrived, he learned that his test results had betrayed him. Though Peking University offered an excellent program in physics, something in Xu's scores led the Ministry of Education to assign him a different fate. "At that time in China, the system was different," he explained. "You did not decide your own major. The government decided." In Xing Xu, they did not see a future physicist. Nor did they think him a likely software engineer, his second choice. It was with a heavy heart that he read his official departmental posting: geology. He felt like dropping out and going home. So began one of the most storied careers in modern paleontology.

"It was not my choice," he told me firmly, "but my family wanted me to stay in Beijing. The only way to remain was to go to the Geology Department and follow the major they gave me: paleontology. It was never my choice!"

I spoke with Xu by telephone from his office at the Chinese Academy of Sciences, where he is now a full professor at the Institute of Vertebrate Paleontology and Paleoanthropology. It's safe to say he has settled into his new role. In the fifteen years since fate and bureaucracy steered him toward a life of fossils, he has named more than thirty new dinosaur species, more than any other living paleontologist. His work appears so often in *Nature*, the top journal in the field, that one colleague suggested they make him a columnist. In the world of feather evolution, his peers all use the same word to describe him: *brilliant*. In the popular press, he's known as China's Indiana Jones.

But it all got off to a rocky start. "For the first two years I did nothing—I ignored my courses," he explained, laughing at the memory of staying in his room, teaching himself computer programming instead. "By the last year I realized I had to learn something and prepare a thesis. I asked myself, 'Maybe dinosaurs are interesting?'"

His timing was perfect. Xu stepped into the world of Chinese paleontology just as it was exploding with new discoveries. Fish, plants, pterosaurs, insects, birds, and dinosaurs poured in from Liaoning Province, where bands of ancient shales alternated with basalt in a group of rocks known as the Yixian Formation. It was the most complete and diverse snapshot of early Cretaceous life ever discovered. "From the beginning I was lucky always to have access to good fossils. Without that I never would have developed a particular interest."

Good is something of an understatement. The fossils from Liaoning are exquisite. Like the limestone at Solnhofen, the fine-grained Yixian rocks preserve amazing levels of detail, from pock-marks in the surface of a bone to veins in the wings of a dragonfly, to that rarest find of all, feathers, lots and lots of feathers. Though dozens of early birds have been unearthed, the real surprise came when feathers began appearing on theropod dinosaurs, much in the way that Huxley or Ostrom might have predicted.

"We expected someday to find primitive feathers, but seeing typical modern feathers on theropods was surprising!" When he talks, Xu's excitement seems to build with each word, as if a new discovery may lie at the end of every sentence. It seems that, having skipped the typical boyhood enthusiasm for dinosaurs, he has developed a full-blown case as an adult. He asks questions and answers them in rapid, accented English, illustrating his points with fossils, many of which he had a hand in discovering. "Do you know *Beipiaosaurus*? Do you know *Caudipteryx*?"

His youth and energy reminded me immediately of Rick Prum, whose first words to me had been, "Sure, I'd be happy to talk with you about feathers. The only problem is, I may never stop talking!" The two share a common enthusiasm but approach the question of feather evolution from slightly different angles—Prum as an ornithologist and Xu as a paleontologist. Together they make a powerful team, and they've collaborated on several occasions. I asked Prum what it was like working with Xu. "He's fantastic, brilliant," came the immediate reply. "I have unlimited respect for the guy." It can't hurt that Xu has an uncanny ability to unearth specimens that support Prum's ideas.

"I have a good field crew," Xu told me modestly, but his knack for finding fossils is legendary. In one famous incident, a documentary film crew asked him to reenact the discovery of a unique sauropod he'd discovered on an earlier expedition. He obliged, returning with them to the dig site and brushing dirt from a random bone while they filmed. As the soil fell away, however, he looked closely and realized it was something entirely new. His random bone turned out to be part of the largest birdlike theropod yet uncovered, a twenty-four-foot, two-ton behemoth that he named, appropriately, *Gigantoraptor*.

Based in part on such serendipitous discoveries, Xu and others have documented examples of the feather stages Rick Prum predicted in a wide range of theropod dinosaurs. I asked Xu to name five fossils that would trace the evolution of the feathers, plotting the path protofeather to feather, from theropod to bird. He quickly rattled off eight names. I told him that my book had a word limit, and he reluctantly narrowed the field to the five remarkable creatures described below. Their stories not only bolster theories of feather evolution but also help answer many of the questions left open by *Archaeopteryx* and give us tantalizing glimpses of the ancient world where feathers first arose.

Liaoning Province lies in far northeastern China, a land of rural villages and heavily industrialized cities also known as Manchuria. In the seventeenth century, when the Manchus conquered the

Mings to found the great Qing dynasty, it's no surprise that they kept the capital in Beijing and moved south. Described by various travelers as "gray," "dusty," or "very brown," Liaoning famously alternates between sweltering summer monsoons and frigid winters. In between, locals eke a living out of the rocky plains and valleys by growing sorghum and fruit or, in more recent years, by digging fossils.

At the dawn of the Cretaceous period, Liaoning looked very different. Forests and lakes covered the rolling landscape, and a range of nearby volcanoes spewed the occasional lava flow or thick cloud of ash. While molten rock doesn't leave much behind to fossilize, ash eruptions can be a paleontologist's best friend. They're often accompanied by waves of heat or gas that kill animals quickly, just in time to be covered by the ash fall. Repeated over eons, such eruptions layered the shallow lake bottoms of Liaoning with fine, powdery ash, providing an ideal environment for fossils.

I've tried this at home, and it works. In winter, a high water table floods the shallow pit behind our house where I dump the stove ash. We heat and cook with wood, so I find myself out there at least once a week, watching another bucketful sink and settle slowly to the bottom. My fossil experiment involved two chicken feathers tacked down and buried in ash, layer after layer, week after week, all winter long. When I returned in August, the pit had dried to a smooth gray hardpan, peppered with bits of charcoal. Even though I'd flagged the locations of the feathers, they were hard to pinpoint in all that chalky grit. And even though I tried to be careful, the feathers came out in pieces. It gave me a sudden appreciation for the challenges of archaeological fieldwork, and I decided the next time I tried to make a fossil, I would excavate with fine brushes and dental picks instead of a garden shovel.

The feathers themselves had started to decompose, but the impressions they left behind were beautiful. In the sunlight I could see mirror prints of each rachis and barb etched into the clods of ash and tinted a slight brown against the gray. These "protofossils" formed after only a few months, under a few inches of sediment. If I'd kept at it for a million years or so, remnants of those feathers might have truly fossilized as the ash turned to paper shale, the very rock type from which a Chinese farmer extracted the first feathered Yixian specimen.

It was 1996 and Xing Xu was still in graduate school when Qiang Ji, his future colleague at the Chinese Academy, purchased a curious-looking fossil from a farmer near the village of Sihetun. If Richard Owen got a good deal on *Archaeopteryx* for £700, Ji's acquisition of *Sinosauropteryx prima* for $750 was a steal. Though his specialty was crustaceans, Ji still knew enough about theropods to suspect that the dark filaments fringing its head, back, and tail might indeed be feathers. He published a description in a Chinese journal, but the real fuss began when pictures of the specimen circulated at that year's Society of Vertebrate Paleontology meeting in New York. People ignored the presentations, crowding in the hallway to get a glimpse of *Sinosauropteryx* and argue about its implications.

Supporters of John Ostrom's theropods-to-birds theory immediately claimed these "protofeathers" as their missing link. Members of the BAND derided them as merely degraded collagen fibers, the kind of common tissue found in shark fins or the crest of an iguana. Though well preserved, the filaments left room for argument. Most observers agreed they resembled Prum's Stage II feathers, but final confirmation came only after electron microscopy revealed details of structural pigmentation. Collagen occurs under the skin and is never colored, but *Sinosauropteryx*, it turns out, sported a bright coat of russet and ginger feathers and a striped tail.

Sinosauropteryx prima, the first feathered dinosaur from the Yixian Formation.

As additional specimens came to light, a surprisingly detailed image of *Sinosauropteryx* ecology emerged. Perfectly preserved stomach contents revealed that it fed on lizards and tiny primitive mammals. Like its close relative *Compsognathus*, it was no bigger than a chicken and was entirely terrestrial—a small fleet-footed, bipedal predator. In this context, its feathers had no aerodynamic function. They may have served as insulation, but the confirmation of color and striping strongly suggests they played a role in display.

Two years later, Qiang Ji and colleagues described another Yixian fossil, *Caudipteryx zoui*, whose feathers showed a leap in development. Clustered on the hands and at the tip of the tail, each featured a distinct rachis and symmetrical vanes. Additionally, the body was covered in dino fuzz. Prum's theory now had Stage II and III confirmed in a single theropod, this one a larger turkey-size species that fed on plant material as well as meat. Several

Caudipteryx zoui, pictured with its feathers fanned in a hypothetical courtship display.

Caudipteryx specimens have included a telling group of small seed-crushing stones, known as "gastroliths," in the exact position of a modern bird's gizzard. Though its wing and tail feathers were vaned, they lacked the asymmetry associated with flight, which indicates that they were primarily for display.

The growing fame of the Yixian Formation and its feathered dinosaurs soon set off a wave of fossil prospecting in Liaoning. Mark Norell, curator at the American Museum of Natural History, describes those heady days with an insider's perspective in his book, *Unearthing the Dragon*. He recalls visiting a regional museum where fossil preparers chipped away with rusty nails and priceless specimens lay stacked up for sale in the gift shop. To local farmers, finding a good fossil could double their yearly income, and it seemed everyone in the area regarded feathered dinosaurs with growing pride. A bas-relief *Sinosauropteryx* even appeared on commemorative bottles of the local hooch, *baijiu*, a strong sorghum-based whiskey.

Xing Xu entered the feathered dinosaur race with a splash while still a graduate student, describing two pivotal specimens in 1999 and 2000. *Beipiaosaurus inexpectus* took its scientific name directly from the unexpected nature of its features. Its simple dentition suggested herbivory, but its size and huge claws pointed to a predatory habit. In reconstructions and artists' concepts, it resembles a sort of hulking, feathered sloth. Even stranger to Xu were broad filamentous feathers that appeared at odd intervals along its back, growing out from a covering of more typical dino fuzz. They are unbranched and probably represent the simple quills predicted as a Stage I feather by Prum. "My other idea," Xu told me in one conversation, "is they could be ribbonlike feathers, created by a partial fusion of primitive barbs!" Either way, the diversity of theropod feather types was expanding.

Coming on the heels of *Beipiaosaurus*, Xu's *Microraptor zhaoianus* created a stir twice, the first time as part of a famous scandal. In 1999, *National Geographic* announced a sensational new Chinese specimen with a toothed beak and fully feathered wings and tail.

Beipiaosaurus inexpectus, a Yixian dinosaur with broad, filamentous feathers along its back and tail.

They called it *Archaeoraptor* and hailed it as the long-sought link between terrestrial theropods and flying birds. The fossil sounded too good to be true, and it was. Soon after examining the specimen, Xu and several colleagues exposed it as a chimera, a composite of several different fossils cobbled together on one slab. The fraud proved a great embarrassment for the National Geographic Society and the paleontological community in general. But, as Mark Norell pointed out, it actually demonstrated the efficacy of the peer-review process; it was only too bad that the magazine hadn't waited for review *before* publishing the description.

To Xu, however, *Archaeoraptor* turned out to be a great opportunity. By returning to the quarry site and talking to local fossil dealers, he managed to reunite the tail of *Archaeoraptor* with the original fossil from which it was taken, and the resulting specimen was nearly as remarkable as the fake.

"*Microraptor* was the biggest surprise," Xu said. "It was clearly a theropod, but with asymmetrical flight feathers on wings and legs, even its feet. It was basically a four-winged theropod, like a biplane!"

Later analyses confirmed that *Microraptor* probably glided from tree to tree, a habit with great implications for the origins of feathered flight. We'll return to the controversy surrounding that topic in Chapter 7. In terms of feather evolution, the fossil confirmed that theropods had asymmetrical feathers (Prum's Stage V), while their apparent flight capacity and herringbone vane pattern also strongly suggested the interlocking barbules of Stage IV. With this fossil, all five developmental novelties could now be argued for in theropods, seeming to close the book on feather evolution. But a final hurdle remained. BAND members called it the "temporal paradox."

Historically, paleontologists dated rock formations by finding correlations between similar groups of fossil organisms, but they can now also measure the ratio of isotopes that slowly decay at predictable rates. Both of these techniques put the Yixian rocks in the early Cretaceous period, 110 to 130 million years ago. *Archaeopteryx*, on the other hand, lived in the late Jurassic, 140 to 150 million years ago. This disjunction spurred critics to ask how the "first bird" evolved from theropods if the feathered dinosaurs occurred tens of millions of years later.

From an evolutionary standpoint, the "temporal paradox" is not a paradox at all, but instead an example of a common pitfall when thinking about evolution: the idea of evolutionary "progress." Everyone has seen that classic cartoon of a parade of life walking from the water up some primordial beach, from fish to reptile to mammal to man. The image has become a powerful intellectual attractor, a notion so appealing that our minds return to it again and again. Unfortunately, that's just not how evolution works. In practice, evolution is less about straight lines than it is about networks, webs of descent where the development of traits is anything but unidirectional. Though there is a general trend from simplest forms of early life to more complex organisms, complexity is not an evolutionary virtue in itself. It develops and persists only when it confers advantage (or at least no disadvantage), and examples abound of complex structures, including feathers, becoming simplified or eliminated over time. In evolution, age alone does not equal ancestry, and apparently simpler forms can readily exist long after more complex ones evolved.

No one proposes that *Sinosauropteryx*, *Microraptor*, or even *Archaeopteryx* was the specific ancestor of modern birds. Rather, their feathers and skeletal similarities suggest that these creatures and birds all shared a common ancestor somewhere deeper in the theropod lineage. Though birds were the only ones to survive, there was certainly a long time when many closely related groups persisted side by side on separate evolutionary paths. The evolutionary paradigm does not call for discrete replacements, one form changing to another and another beyond that. It's a far messier and more glorious accumulation and elimination of varieties.

The next time you chance upon a dog show on TV, or tour the local animal shelter, remember that all those myriad varieties, from schnauzers to setters, developed quite recently through selective breeding. Every dog in the world descends from gray wolves first domesticated in central Asia between fifteen and thirty thousand years ago. Since then they have traveled with people to every continent, and countless breeds have come and gone. The Pekingese dates to the courts of imperial China more than two thousand years ago, the paisley terrier became common in Victorian England but died out in the 1920s, while the now popular labradoodle was developed in Australia fewer than thirty years ago. Meanwhile, the gray wolf persists in the wild in its original, unaltered form. To future paleontologists, dealing with only a few scattered fossils, this will be an unholy mess to make sense of. They would be totally incorrect, for example, to assume that the Empress Dowager's Pekingese was the ancestor of a modern wolf, simply because it existed earlier in time. They would be right, however, to conclude that all dog breeds and the modern wolf share a common ancestor somewhere earlier in the canine lineage.

Though this argument makes a sound defense, proponents of the theropod-bird story still longed to find a feathered dinosaur older than *Archaeopteryx*, one that would put the temporal paradox to rest once and for all. "If you want certain fossils, you have to prospect in sediments of the right age," Xu explained simply, as if anyone with a shovel and a geologic map could replicate his most dramatic discovery to date.

Following his own advice, Xu and his team found *Anchiornis huxleyi* by delving below the Yixian shales to the little-known Jurassic rocks of the Haifangou Formation. Like *Archaeopteryx*, *Anchiornis* sported distinct asymmetrical feathers and may have been capable of gliding flight, but this creature lived more than 160 million years ago, at least 10 million years before the "first bird." It was a small dinosaur, only a foot long, and resembled *Microraptor*, with downy body feathers as well as feathered wings, legs, and feet. *Anchiornis* also sported a distinct feathered head crest, not unlike a modern Cardinal or a Steller's Jay. Named in honor of Thomas Huxley, this fossil was presented by Xu at the 2009 Society of Vertebrate Paleontology meeting, neatly tying up one of the last serious criticisms of the theropod-bird link that Huxley suggested nearly 150 years before.

Taken together, these feathered dinosaurs (and the twenty other species uncovered to date) make a pretty compelling case for Prum's developmental theory. They all bear Stage II filaments; *Beipiaosaurus* has Stage I quills. *Caudipteryx* covers Stage III (and possibly IV), while *Microraptor* and *Anchiornis* take care of Stages IV and V. "We can't see everything," Prum admits. "Barbules are pretty small to fossilize, and it's hard to tell when things are hollow. But overall I'm pretty pleased with what the fossil evidence tells us."

The fact that all five feather stages can be found in theropods as well as in modern birds underscores the close relationship between these groups. It adds crucial evidence to what may be the first real scientific consensus on avian and feather origins. Plenty of details remain to haggle over, but the vast majority of paleontologists and ornithologists now find the theropod-to-bird framework convincing. Kevin Padian, an evolutionary biologist from the University of California at Berkeley, put it this way: "The earth is round, the sun doesn't go around it, the continents move, and birds evolved from dinosaurs."

No two scientists have done more to advance current ideas of feather evolution and avian origins than Prum and Xu, but even they don't agree on everything. In conversation, both struck me as intensely curious, constantly questioning, refining, and challenging their own ideas as well as each other's. Though I never talked with the two together, or even on the same continent, I kept a sort of conversation going through phone calls and e-mails and when I met Prum in his office at Yale's Peabody Museum of Natural History. The points on which they agree frame our understanding of feather origins, but their differences taught me something about the nuances of evolution itself.

A few years after Prum's theory came out, Xu published his own model of feather development. It's similar but allows for quills and filaments (Prum's Stages I and II) to develop before the invention of the follicle. Xu feels this modification better accounts for what he sees in the fossils, where evidence of integumentary filaments, or "dino fuzz," is becoming more and more widespread. "This year we published a new species of Ornithischian [a group of dinosaurs that includes *Triceratops* and other armored species only distantly related to theropods] with long filaments that could be protofeathers," he told me. Something resembling dino fuzz has even shown up on pterosaurs, and Xu thinks that filaments may have originated before theropods, or may have evolved independently many times.

"That's possible," Prum admitted, when I asked him about Xu's idea. "Our own research more or less came to the same conclusion. But you wouldn't have had any structural complexity until the follicle—no barbs, no rachis. Without the follicle, a feather would basically be like a wart."

In Xu's thinking, the early stages of feather evolution were extremely unstable. Quills, filaments, or possibly even more advanced forms could have appeared and winked out again many times in different lineages. "Novel structures can easily disappear and reappear," he explained. "Only after they stabilized could feathers diversify in form and function. Maybe that stabilization only occurred sometime around *Anchiornis* or *Archaeopteryx*."

Prum and Xu both agreed that feather evolution was iterative, an accumulation of novel morphological traits. Then when, I asked, do you call a feather a feather?

"If it's a hollow tube, it's a feather," Prum replied immediately. "One thing I keep saying again and again is that there's no such thing as a 'protofeather.' No one talks about a 'protolimb.' You either have a limb or you don't. Why should feathers be any different? If it's a tube, it's a feather. Period."

Xu was more equivocal. "I've been thinking a lot about this. How do you define a structure? These days everyone agrees that taxonomic names are arbitrary. What is a dinosaur and what is a bird—if one evolved gradually into the other, where do you draw the line? I argue it's the same with structures." He paused for breath, and then more ideas tumbled out. "Feathers have unique morphology, chemistry, keratins, structural features, et cetera. It's very likely that those complex features evolved in a steplike way, sequentially. So where do you draw the line between feathers and nonfeathers? I don't know."

Alan Feduccia has a different perspective. In spite of the growing theropod consensus (or perhaps in some ways because of it), the remaining members of the BAND continue to publish critiques, questioning methods and results and looking for holes in the evidence. Feduccia doesn't deny that fossils like *Caudipteryx* or *Microraptor* have feathers, but he prefers to consider them secondarily flightless birds, like modern ostriches, rheas, or emus. "You can boil my whole position down to one statement," he told me: "If it has bird feathers, it's a bird."

"That's their latest ploy!" Prum exclaimed when I told him this. "If they can't deny that the dinosaur's feathers are real, then they just call it a bird. And that's after saying for years that these same fossils weren't even related to birds!" He admits a certain exasperation with Feduccia and company, who criticize his ideas without offering any clear, testable alternatives. In fact, continual reporting on the BAND's minority viewpoint (in the interest of journalistic fairness) may be perpetuating a controversy that most scientists consider over and done with. But when I pressed him, Prum did admit that certain criticisms had helped refine his thinking, whatever the cost to his blood pressure. "This whole debate is really overblown," Feduccia told me. "Everyone agrees that birds and dinosaurs are related. Prum's group thinks they evolved from theropods. My group just thinks they evolved much earlier and remained distinct from dinosaurs."

Feduccia has long argued that birds evolved from an as yet undiscovered archosaur, one of the ancient reptiles that predated and gave rise to the dinosaurs (as well as the pterosaurs and the crocodilians, the latter of which include the modern reptiles). In the BAND's version of the story, birds and theropods are more like distant second cousins. Their host of similar traits came about less through relatedness than convergent evolution—similar adaptations to suit similar lifestyles. He illustrates his point with a skeletal feature of birds and theropods that doesn't appear to match. It's one of the last lingering questions in the theropod-bird theory, an apparent inconsistency in which digits make up their three-fingered hands. Since both evolved from a five-fingered condition, the question is, which two digits were lost? In birds it looks like one and five, while in theropods it's four and five.

Prum and Xu are not convinced. They counter that developmental and molecular studies show that digit identity is malleable and that various developmental patterns could be tweaked to a three-fingered result. Xu allowed that "settling the digit issue is definitely a research priority" and pointed to a promising new theropod fossil whose first digit was dramatically reduced, a possible analog to the pattern in birds. Either way, he and Prum consider digits a minor snag in the face of overwhelming evidence.

Feduccia is fond of countering the theropod "orthodoxy" with a line from H. L. Mencken: "For every complex problem there is a solution, which is simple, neat, and wrong." As fossils, studies, and expert opinions line up against him, however, his own position is starting to sound like the simple one. At the end of our conversation, I asked him what it felt like to be constantly swimming against the tide. "Look, I don't know how all of this will play out," he answered with a sort of weary good cheer, "but I do know that the orthodox viewpoint will be challenged on all kinds of levels."

Feduccia is certainly right that the orthodoxy will be challenged, altered, and refined as new fossils come to light—that's just how science is supposed to work. But perhaps for the first time, the theoretical framework appears robust, and the arguing will now be about the details. When I asked Xu what the weak points were in the theropod and feather development models, he had to stop and think. It was the longest period of silence in our conversation. "No, there is no real weakness," he said finally. "The only thing we need is more evidence. We have this big framework, but there are so many details to fill in." It sounds like it's still a good time to be a fossil hunter.

Rick Prum gave me a similar response. "If I had to alter anything from my original paper, I might de-emphasize the importance of the follicle," he said, but otherwise the model was working well and leading in all kinds of interesting directions. He compared feather research to stumbling across a new, unexplored valley of science. "You top a rise and suddenly there's this beautiful place with a braided river running down the middle, and you hike into it thinking, 'No one has ever been here before!'"

I left Prum's office with a mind full of possibilities; every answer he'd given me seemed to spawn some new and interesting question. If theropods were feathered, how birdlike was their behavior? If they were brightly adorned, then how and when did color evolve? If feathers have been around for so long, what strange shapes and functions might have come and gone? My visit to Yale coincided with a heavy snowfall, and later that same day I took a train to New York City and found myself walking through a wintry scene in Central Park. Head still swimming, I hardly noticed the theropod descendants flitting and flapping through the trees above me until I spotted a small dark feather lying atop the fresh snow.

I picked it up, its perfect, symmetrical vanes just the right shade of bluish gray for a Common Pigeon, that undisputed sovereign of city birds. After my day with Prum, I couldn't help but visualize the feather developing, emerging from its follicle in a process the model describes with phrases like *proliferation and outfolding of the basal layer of the feather germ epithelium*. What science makes complicated, the birds carry off with ease, growing new feathers, sloughing the old, staying warm, keeping cool, and even changing colors with the seasons.

That lone plume reminded me there were plenty of other feather topics for science to make complicated. Why do birds molt, and what triggers it? Why molt so often? How and where do the feathers come in? How do the same follicles produce such vastly different colors and shapes? Just the sort of questions a little muttonbirding could answer.

A Common Pigeon in flight above its urban habitat.

CHAPTER FOUR

How to Catch a Muttonbird

A little cock sparrow sat on a green tree,
And he chirruped, he chirruped, so merry was he.
A naughty boy came with his wee bow and arrow,
Says he, I will shoot this little cock sparrow;
His body will make me a nice little stew,
And his giblets will make me a little pie too.
Oh, no, said the sparrow, I won't make a stew,
So he clapped his wings and away he flew.

—Mother Goose traditional rhyme

I stretched my hand into the darkness and felt something feathered scuttle away, just beyond my fingertips. Lying on my stomach in wet, guano-streaked grass, I had my arm shoulder-deep in a narrow hole that reeked of fish oil. I reached again but felt only mud. This was as close as I would ever get to muttonbirding, and, frankly, it was something of a relief to fail.

Moments earlier I had watched seabird expert Peter Harrison enthusiastically yank a young Thin-billed Prion from its burrow.

He sprang up from the ground and flourished the little bird like a magician pulling a bouquet from his sleeve. As we crowded around for a look, the prion blinked a few times at the daylight, but otherwise sat placidly and unconcerned in Peter's hand. On New Island at the western tip of the Falkland archipelago, prions had learned to fear falcons, skuas, rats, and the occasional feral cat, but a group of bird-watchers and a British ornithologist were nothing to worry about.

Recently knighted for his contributions to science and conservation, Sir Peter is reportedly the only person in history to have seen every species of oceangoing bird on the planet. His illustrated field guide, *Seabirds*, is considered definitive, and whenever he leads a tour, avid birders and even professional ornithologists snatch up the tickets like Grateful Dead fans vying for a front-row seat.

Not surprisingly, Peter handled the prion with practiced ease, pointing out the distinctive hornlike cylinder at the base of its bill that put this bird in the family Procellariidae, "the tube-nosed swimmers." "Their sense of smell is so keen they can find their burrows at night and sniff out patches of krill on the open sea," he said with a note of wonder, as if, like the rest of us, he too was seeing a prion for the very first time.

Barely more than a hatchling, the plump, fist-size bird he was holding would soon exceed its parents in weight, packing on the layers of fatty energy it would need to transform itself from a fuzzy gray ball to a sleek flying machine. In less than sixty days, more than two million young prions would take their first flight through the skies over New Island and then disappear out to sea. As the long austral winter set in, the birds would wander alone or in small flocks, searching constantly for food. They wouldn't touch ground again for nearly a year.

With their combination of rich, oily meat and utter helplessness on land, it's no surprise that the chicks of prions and their

A Thin-billed Prion chick on New Island.

larger relatives, petrels and shearwaters, once featured prominently in the diets of whalers, fishermen, and anyone else within striking distance of a nesting colony—not just in the Falklands but throughout the southern oceans. The name *muttonbird* dates to the eighteenth century, when lonely sailors assured one another that grilled procellarians smelled and tasted just like the roast mutton from home. Having stuck my nose into a prion burrow, I will put that notion down to wishful thinking. But what muttonbirders lacked in epicurean precision, they more than made up for with an intimate knowledge of their quarry: habitat preference, breeding behavior, and the precise timing of feather growth.

After a brief inspection and a few photos, Peter returned the prion chick to its burrow, being careful not to plunge it into the wrong nest hole and start a territorial ruckus. Later, when the rest of our company dispersed to explore the island, I knew I had to try my hand at muttonbirding. Finding another nest was not a problem. With literally millions of burrows on the island, the ground felt spongy underfoot, and every step triggered eerily disembodied clicks and hisses from the startled birds below. Some tunnels still held adults incubating their eggs, but most now contained a single downy chick. From soon after hatching to the moment they fledged, the chicks waited out the daylight hours alone, their parents returning to feed them only at night, when they would be safe from predators under the cover of darkness.

The burrow I chose turned out to be just longer than my arm could reach, and the chick it contained simply scrambled out of my grasp. In a way, that suited me fine. I can't deny that I have an urge to capture, poke, and prod everything in sight, but such behavior is sometimes at odds with a deeper conservation ethic. The goals of science may be noble, but there's no avoiding the fact that the practice of field biology can be terribly impolite to its subjects. The prions underfoot seemed innumerable, and I could have pressed on to search more burrows, but I didn't want to make the all-too-common error of mistaking abundance for resilience. Failing to catch my muttonbird allowed me to satisfy the urge to try, and then get on with the business of looking about the island in general—soaking up the euphoric reek and clamor of all that concentrated bird life.

Black-browed Albatross and King Shags wheeled and flapped overhead, passing low across the hilltops en route to their nests. Gentoo, Rockhopper, and Magellanic Penguins trudged between the beaches and their grassy rookeries, pausing to lift their bills skyward and call out in long, honking ululations. Among the tussock grass, I caught glimpses of Snowy Sheathbills striding about, bent forward like tiny professors lost in thought. Kelp Gulls, Giant Petrels, and Brown Skuas lurked everywhere, hoping to steal eggs or hatchlings from among the island's forty nesting species, or to surprise the odd prion that might venture a daylight landing near its nest hole.

With slight variations, this chaotic scene plays out every year on islands throughout the southern oceans, a defining backdrop to the business of muttonbirding. And a business it is, with more than 250,000 Sooty Shearwaters and up to 150,000 Short-tailed Shearwaters still taken annually in the islands off New Zealand and Tasmania. Plucked, cleaned, and salted, a brace of plump muttonbirds fetches as much as twenty dollars in local markets, and even more through online sales. The trade remains an important income source for scores of Maori and Aboriginal families, who have perfected their hunting practices to a fine degree. An experienced muttonbirder can locate, snatch, and bag the occupant of a burrow every 5.6 minutes and might net a cool thirty thousand dollars in thirty days of hunting.

The Maori divide their muttonbirding into two distinct phases based on precise knowledge of their quarry's natural history. My hapless attempt occurred during what they would have called the *nanao*, the period of daylight gathering when plump chicks are snatched straight from their burrows. This captures the birds at their fattest, and *nanao* chicks command the highest prices. Nighttime gathering is known as the *rama* and occurs on select evenings toward the end of the season, when the young birds emerge from their burrows en masse and teem across the grasslands toward high points or cliffs from which they will take their first flight.

Muttonbirders succeed through a simple but profound understanding of feather biology. What happens between the *nanao* and the *rama* is the muttonbirds' first molt, a complete transformation from natal fluff to the diverse array of contour feathers, flight feathers, semiplumes, down, bristles, powderdown, and filoplumes that will see the young birds through their first year at sea. *Nanao* birds are helpless chicks who wouldn't survive a rain shower; *rama* birds are protected by plumage that can withstand some of the harshest weather on the planet. Growing that feather coat in a matter of weeks requires a massive investment of energy, burning quickly through the chicks' substantial fat reserves. (Gorged on fish oil and regurgitated krill, some youngsters get so large before the molt that they can't be extracted from their burrows, but all emerge during *rama* as sleek, powerful fliers.)

Field scientists learn about biology through data analysis and long, slow hours of measurements, note taking, and other small attentions. An older and more direct route to biological knowledge lies through the stomach. Subsistence hunters gain insights quickly or suffer the immediate consequences. I once sat in a deer blind with a young man whose family had relied on hunting, their quarry changing with the season. As the long, cold hours passed by without a deer, he made sport by naming every duck that flew overhead, recognizing each species by the particular whistling of its wings in the cold morning air. I had binoculars with me and checked the birds as they passed. He was right every time.

Catching muttonbirds provides a good illustration of early feather growth, but it's also a prime example of "essential knowledge." Hunters need to know certain things about their quarry or they go hungry, but they don't need to know everything. For a muttonbirder, it's essential to know when and where the birds breed and molt, but you could still fill your *nanao* sack without studying how their chromosomes pair up during cell division. This is an important lesson to keep in mind as we explore the intricacies of feather structure and development. To truly appreciate feathers, we should understand the basics: what they're made of, how and why they molt, and just how a follicle produces such variety in shape and form. But the chemistry, physics, and molecular genetics underlying these processes are deep and wide pools that continue to challenge career scientists. Like muttonbirders, we will restrict our hunt to the essential points.

Though we don't think of feathers in the same way as bacon, steak, or a ham sandwich, they too are a good source of protein. From the quill to the rachis to the vane, a feather's parts consist largely of keratin, the same kind of protein found in hair and fingernails. People don't generally eat feathers, but their nutritional potential has not been overlooked by the animal feed industry. Chicken and turkey processors in the United States churn out more than ten billion pounds of feather waste every year. They turn a tidy profit by channeling these pluckings to the likes of ConAgra and Purina, where the feathers are boiled, dried, and ground into a protein-rich meal that finds its way into everything from canned dog food to cattle pellets. In a macabre twist, it's even fed to chickens. Feather keratin also plays a role in organic farming, where feather-meal fertilizer is considered a natural way to boost soil nitrogen. Applied to rows of lettuce, beans, or other nitrogen-loving veggies, it provides a surprising pathway for protein from feathers to reach the organic produce aisle.

I once took biochemistry from a glowering professor with a furrowed brow, wild hair, and a thick Polish accent. He began his lecture on keratin by showing us a close-up photo of a rhinoceros. "Rhinoceros," he thundered at the class, "is like tank!" He could just as easily have shown a slide of a song sparrow or a warbler, but it wouldn't have had quite the same effect.

What he meant was that keratin is a protein designed for strength. Its long molecules form tough, fibrous webs that bulk up all kinds of body coverings, from rhino hide to turtle shells. Feathers are full of it, and so are fingernails, scales, hooves, claws, horns, and hair. It evolved early, before the first vertebrates emerged from the sea, and now constitutes a basic structural building block throughout the animal kingdom. In that time keratin has become exceedingly common, but not all keratin is created equal.

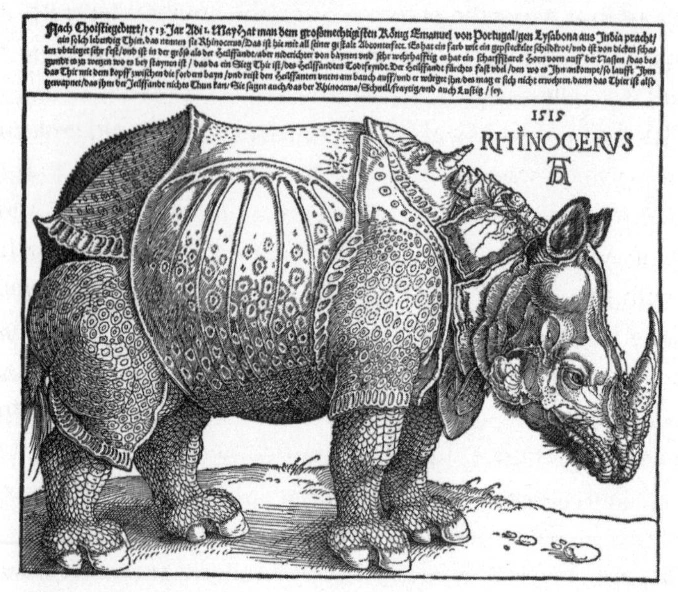

The armorlike hide of a rhinoceros contains large amounts of keratin, a protein designed for strength.

In the opening sequence of *The Graduate*, a family friend takes young Benjamin Braddock aside and gives him one word of career advice: "Plastics." Benjamin's awkward, stone-faced reaction sets the tone for the whole movie, but the important point here is the use of the plural. To the uninitiated, plastic is plastic. But this fellow knew there are scores of varieties, each specially formulated to its particular purpose. Recyclers separate them into basic polymer types, identified by the numbers you see stamped on containers of all kinds. Soda-pop bottles bear the number 1 and can be recycled into new bottles or a variety of synthetic fabrics. Milk jugs are number 2s and get turned into plastic lumber and deck furniture. Plastic bags are number 4, disposable coffee-cup lids are number 6, and so on. They're all plastic, but subtle variations in their molecules make each type distinct, differences that become obvious in the recycling process. Keep the numbers separate, and you get valuable plastic for new products. Mix them up, and the result is a useless slag.

It's the same story with keratin. Like plastic, keratin molecules form polymers where subtle variations in shape and composition produce distinct differences in the final product. If there were a keratin recycling program, you would never mix the owl feathers in with the monkey fur or goat hooves because feather keratins belong to a group utterly distinct from those found in mammals. They're closely related to reptile keratins, and the details encoded in their genes form part of the argument that birds evolved from dinosaurs. Genetically and chemically, feather keratins stand alone, perfectly suited to their purpose. They provide the molecular basis for many important feather characteristics: strong yet light, firm yet flexible, durable, colorfast, and elastic. You could not build a feather from cow horn—it's too brittle. Eyelash keratin is too soft, and fingernails tear too easily. In short, feathers are feathers, hooves are hooves, and never the twain shall meet.

Of course, a bird gives no more thought to its keratin than we do to ours when brushing our hair or trimming our toenails. We care only for the appearance and functionality of the end result, and so does the bird. If I had crawled inside that muttonbird burrow to watch my Thin-billed Prion grow, I would have seen the preening instinct start early, as soon as the first pennaceous feathers appeared. Even in the darkness of its nest hole, the young bird would have devoted hours to plumage care, grooming each feather repeatedly in anticipation of its maiden flight and the ravages of wind, rain, and waves that would soon frame its daily life. Among the first tasks: clipping away fuzzy remnants of natal down that still clung to the tips of each incoming feather.

"You can think of every feather produced by a particular follicle as part of one long, continuous tube," Rick Prum told me.

Starting with natal down and moving through juvenile, adult, and breeding plumages, each individual follicle produces a range of feather types and colors over the lifetime of a bird. They're programmed to grow their feathers rapidly and then turn off, resting for months or even a full year until a new feather is needed. Then the follicle comes back to life, pumping out keratin again in the intricate helical pattern that helped Rick define his evolutionary model. And just as natal down clings to the end of the first juvenile feathers, every subsequent adult feather is connected too, however briefly. Examine the base of any shed plume, and you will find a tiny hole, the umbilicus, where that feather once adjoined its successor before being pushed aside as the new plume emerged from the follicle.

This is molting, an annual or semiannual process of renewal that helps define the life cycle of birds. It relies on one of Rick Prum's key evolutionary novelties: the feather follicle, a biological marvel capable of intricate variations, from a bird's first down to the juvenile plumage, to adult feathers that often differ dramatically between breeding and nonbreeding seasons. Each follicle is a tiny cylindrical depression in the skin, surrounded by muscles and nerves and with a central core of live tissue. Follicles foster structural complexity in part by nourishing the feather as it grows. Unlike hair, which emerges as a simple string of dead cells, a developing feather, known as a pinfeather, grows around its live core and will bleed profusely if damaged. Its ultimate structure—the specific arrangement of barbs, barbules, rachis, and so on—is determined in the follicle collar, and that growth pattern proceeds in a helical fashion as the feather emerges. Pinfeathers remain firmly rooted in live tissue until they mature. (*The Joy of Cooking* suggests removing them with pliers and the tip of a sharp knife.) Only then do the blood vessels retract to leave the feather's familiar hollow quill.

When I defrosted and plucked my Winter Wren, several aspects of its feather follicles became immediately clear. Though the feathers appeared evenly arranged, I noticed that they came out in tufts, their shafts concentrated in certain regions. Like most birds, the wren's follicles occurred in well-defined tracts that ran down the spine and rump, along the sides and wings, and in other key areas from which they spread across the body to provide full coverage. Ornithologists still debate the purpose of these tracts, but most believe that clustering feather growth offers two advantages. It distributes plumage across the entire body while allowing skin between the tracts to remain relatively bare, an important consideration for regulating body temperature, as we'll see in the chapters ahead. The tracts may also play a role in how feathers move, helping to concentrate the relevant muscles in discrete lines.

Though my wren's body was no bigger than a matchbox, some of its feathers proved surprisingly difficult to yank free. Each follicle is surrounded by strong muscles and nerves that give birds surprising agility with individual feathers. They can fluff them for warmth, lift them for preening or display, and even make fine adjustments during flight to maximize aerodynamic efficiency. Certain feathers also act as critical sensory organs, particularly facial bristles and the tiny filoplumes that surround flight feathers. Considering the thousands of feathers that cover even the smallest bird's body, coordinating such movements is quite an engineering feat. It would be like a person straightening their part with a thought, twitching individual ear hairs, or accurately judging wind speed from the play of a breeze across their eyebrows.

For most birds, the molt takes place over a period of weeks or months, a gradual replacement that never leaves any one area overly exposed. The molting of flight feathers in particular must be staggered, and they typically drop in sequence from the innermost primaries outward to the wing tips. Watch for this in hawks or other large soaring birds when they pass overhead, and you'll often see symmetrical gaps mirrored in opposite wings, where molting feathers create narrow windows to the sky above. In some species, the molt takes place more precipitously and can leave them flightless until the new feathers emerge. Ducks molt in this fashion, and hunting them during these periods of helplessness helped give rise to the phrase "a sitting duck."

The molt can indeed be a precarious time. As any chicken owner knows, growing new feathers takes a huge investment of energy. Our hens do it once a year, a frustrating period of months where we find ourselves feeding four healthy layers every day yet buying our eggs at the grocery store. The birds can't help it—in the ordered priorities of survival, maintaining one's feathers takes clear precedence over egg production.

Given the risks and costs involved, why should birds bother molting at all? Why don't wrens, muttonbirds, and chickens simply grow a fine set of feathers straight off and stick with them? From a behavioral standpoint, the answer lies partly in breeding strategies. Most birds are inherently visual creatures, and they often recognize and judge potential mates based on seasonal changes in plumage. Females stand to benefit if they can rank potential suitors, separating out the desirable adult males from inexperienced juveniles or yearlings. Seasonal molts allow birds, particularly males, to advertise their status during breeding but remain less conspicuous during the off-season. Over evolutionary time, this system has produced wild extremes in color and display, a topic we'll explore thoroughly in later chapters.

Breeding colors aside, birds molt for the same simple reason that cars need new windshield wipers, guitars need new strings, and biologists need new field trousers. Like any equipment put to constant use, feathers wear out. Sunlight, rain, snow, and the constant friction of flight all take their toll, and even a keratin-tough feather needs to be replaced eventually. The life span of a feather coat relates in part to habitat use and behavior. Soras, rails, and other birds that skulk and scurry through thick vegetation molt twice as often as other species—their feathers erode quickly from the endless abrasions of life in the grass. In some cases, the rate of feather wear coincides neatly with the need for a distinctive breeding plumage. Fresh meadowlark breast feathers appear dull brown when they grow in, but the tips wear away over the course of the winter to expose the brilliant yellow below, just in time for the spring mating season.

In addition to physical wear, however, every feather faces the ceaseless gnawing of a group of voracious hitchhikers called feather lice. Just as Purina knows the value of a feather-meal diet, so too do these tiny insects. But where dog chow comes from poultry waste, lice take their meals when the feathers are still very much in use. Young birds acquire them in the nest, from direct contact with their parents and siblings. For the rest of their lives, they will battle these parasites through a combination of preening, bathing, and vigorous scratching. One group of birds has taken this fight a step further. In New Guinea, the Hooded Pitouhi and its close relatives secrete a powerful neurotoxin that leaves their vibrant orange and black plumes virtually lice free. It's the same compound produced by poison arrow frogs, and pitouhis are the only birds that have evolved a way to make it, presumably by co-opting a chemical from certain beetles in their diet. For the world's other ten thousand birds, however, molting provides the most reliable way to control parasite damage, keeping the lice at bay (and well fed) with a steady supply of fresh plumage.

Under magnification, a feather louse looks like the kind of spiny, segmented space alien that terrorizes people in science fiction movies. Its mandibles appear capable of crushing anything, but in fact are sized to attack only the smallest feather structures—barbules and the fine tips of down. Heavily infested feathers take on a frizzled look as their barbs are stripped clean, losing much of their insulative and waterproofing qualities along the way. Various fungi and bacteria degrade feathers as well, and though birds have developed elaborate ways to combat them (including rubbing themselves with toxic ants, snails, and fruit), only molting can ultimately preserve a feather coat from its parasites.

The final reason to grow new feathers is to replace those lost or broken. When field biologists get to telling stories, the topic often turns to the various "animal bloopers" they've seen—cheetahs peeing into the wind, leaping monkeys missing a branch, or a majestic stag getting his antlers stuck in a shrub. For birds, bloopers often involve crash landings or midair collisions that result in damaged feathers. Physical contests for territories or breeding rights can also lead to injury. In most cases, birds have the ability to relax specific follicles, dropping the broken plumes and triggering a "repair" molt. This same ability can be used for a mass release of feathers in moments of stress or fright. I once found a discouraging pile of chicken feathers in the grass beside our coop, but when I counted heads every hen was there, happily pecking away. The feather pile simply documented a near miss, someone's close encounter with a diving hawk. This adaptation is almost certainly rooted in defense, a way to leave potential predators with nothing more than a feathery mouthful.

Molting serves many functions and appears to have evolved early on, at the same time as feathers themselves or at least with the development of the follicle. One of the oldest known feather fossils is a single *Archaeopteryx* plume, a wing feather apparently dropped in the ancient Solnhofen muck soon after molting. Recently, Xing Xu discovered a series of fossils from a small theropod dinosaur whose juveniles sported drastically different feathers from those found in adults. "This baby dinosaur has bizarre flight feathers," he said in an interview, noting that no modern birds display such a prominent change in feather structure in different molts. Clearly, the complexities of feather growth have been around for a long time.

We now know that feathers are made of keratin, and we understand how and why birds (and dinosaurs) molt, but exactly how the follicle works requires a few more words of explanation. For Rick Prum, his first "aha" moment about feather evolution came when he understood this process.

"The way a branched feather grows is incredible," he said to me more than once, and we watched a computer graphic that he designed to use in his lectures. It pictures the barbs of a vaned feather emerging in single strands from the collar of the follicle and elongating as they move around the rim to fuse with the growing rachis. Imagine people in a crowded sports arena doing "the wave." As it passes, each individual stands up and raises his or her hands at precisely the right moment to keep the wave moving fluidly around the stadium. Cells at the follicle collar behave the same way, but instead of standing and waving, they add keratin to the growing feather barb. It's called helical growth because it progresses like a spiral (or half-spiral) around the rim of the follicle.

The wave analogy works well if you think of standing up as the "on" position (keratin production) and sitting down again as "off" (no keratin production). To make the vaned portion of a feather, keep the cells in one small section of the stadium turned on continuously to form a solid rachis, while the barbs form in symmetrical waves passing down each side of the stands. (If that's hard to visualize, imagine everyone stacking their empty beer cups onto their neighbor's at the instant the wave passes. By the time each wave reached the rachis, the last person in line would be holding a very tall tower of cups—that's a feather barb.) When the vane is finished and it's time to produce the quill, simply have everyone in the stadium stand up at once and a solid tube of keratin will emerge. At the end of the quill, tell them all to sit down, return to their beer and popcorn, and enjoy the game. You won't need their services again until the next molt.

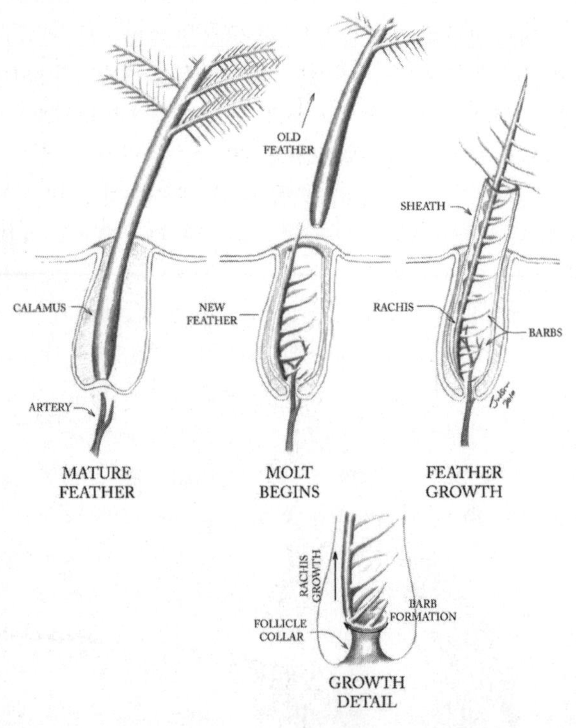

Feather Growth and Molt. On the left, the calamus of a mature feather rests inside its follicle, disconnected from the blood flow of the live tissue below. When the molt begins (middle), the follicle produces barbs and a rachis for the new feather, nourished by live tissue that now extends up inside the arc of the growing barbs. The old feather is pushed out and the new one takes its place, barbs unfurling to form a vane as they emerge from a temporary sheath (right). Feather growth concludes with formation of a solid tube of keratin, the new calamus. In the growth detail, barbs emerge helically from the rim of the follicle collar, like fans doing the wave in a stadium, before fusing with the solid rachis and proceeding upward.

Varying the location and timing of keratin production at the follicle collar can account for every type of feather, from bristles to down, primaries to filoplumes. Even feather oddities begin to make sense in this context. The solid-red wing tips that give waxwings their name are simply fat globs of keratin—the rachis section of the stands stood up for a while on their own before the waves of barb production started. And the bizarre flags on a King of Saxony Bird of Paradise plume grow when alternating sections of the stadium stand up at once, producing sheets of keratin instead of the typical thin barbs. To accomplish these feats, the follicle's cells must act in perfect concert, a symphony of starts and stops that is controlled by a particular gene. It's a famous gene and the only one that we'll talk about in this book. People call it *Sonic Hedgehog.*

Breeding plumage of the male King of Saxony Bird of Paradise.

The scientists who discovered Sonic Hedgehog were studying what controls patterns of growth in all animals. How are the repeated segments of a fruit-fly larvae formed, or for that matter the segments of a backbone or the fingers on a hand? What establishes patterns of top and bottom, front and back or left and right, and how are they maintained as cells grow and are replaced over the lifetime of an organism? What they found was a small family of signaling genes that act like on and off switches, helping tell cells when and how to grow. (In fact, these genes act more like dimmer switches, controlling not just the activation of cell growth but its intensity as well.) Lab technicians dubbed them hedgehogs as early as the 1970s, before they were fully understood, since mutant flies that lacked them produced bristly little hedgehog-like balls instead of normal segmented larvae. Without functional hedgehog genes, cell growth can't be controlled, and there's no way to establish a consistent pattern. Sonic is the most ubiquitous and well studied of the hedgehogs, found in every human cell and in nearly all animals, from hookworms to panthers to birds.

A large team of people worked on isolating and decoding the Sonic Hedgehog gene and their efforts contributed to a Nobel Prize in Physiology in 1995. We know they spent countless hours in the laboratory and that they were meticulous, patient observers. We also know that they played a lot of video games.

To those versed in video gaming, the character Sonic Hedgehog is a legendary figure. Like Mickey Mouse to the cartoon or James Bond to the spy thriller, Sonic helped define a genre, starting off in the early days of gaming and going on to a long, storied career. From arcade machines to Nintendo, Xbox, and the iPod, games featuring Sonic have sold more than seventy million copies. My own experience with video games started in the 1970s with a great affection for Pong but petered out soon after the demise of

Atari. Even so, I had heard of Sonic Hedgehog and wondered about how he became associated with cutting-edge genetics.

In terms of feather growth, the Sonic gene is absolutely essential—after all, few structures in nature are more highly patterned than a feather. When researchers examined the genetics of feather growth they found Sonic Hedgehog directing the show, coordinating that intricate dance of starts and stops. Of course, to call Sonic Hedgehog merely an on-off switch is a gross simplification. It is a gene that produces a protein that acts in concert with other proteins to activate (or inhibit) a highly conserved metabolic pathway—straightforward enough, I suppose, to a molecular geneticist, but I found it very hard to visualize. The stadium wave had helped me intuitively understand the complicated process of helical feather growth. I wondered if playing a video game might do the same thing for feather genetics.

The original Sonic adventure—which, if you're interested, is easy to find on the Internet—takes place in a lurid green landscape of two-dimensional fields, spotted with blocky trees and bright flowers. Its famous namesake is a small blue fellow with a zigzag haircut and red sneakers, who dashes and leaps about, dodging various foes in a manic pursuit of treasure. I thought I was getting the hang of it after I encountered some floating gold rings and managed to evade a diabolical fireball-throwing monkey. But then my hedgehog fell into a deep crevasse, never to be seen again. The experience didn't clear up any of my questions, but it provided a powerful metaphor for the risks of following one's feather obsession too far down the path of developmental genetics. I shut down the computer and turned my attention to something more familiar: Golden-crowned Kinglets, ice storms, and the evolution of fluff.

Fluff

It was the strangest sensation conceivable, floating thus loosely in space, at first indeed horribly strange, and when the horror passed, not disagreeable at all, exceeding restful: indeed, the nearest thing in earthly experience to it that I know is lying on a very thick, soft feather bed.

—H. G. Wells, *The First Men in the Moon* (1901)

CHAPTER FIVE

Keeping Warm

Go tell Aunt Rhody,
Go tell Aunt Rhody,
Go tell Aunt Rhody,
The old grey goose is dead.
The one she's been saving,
The one she's been saving,
The one she's been saving,
To make a feather bed.

—from "Aunt Rhody,"
traditional folk song

A sudden flurry of whisper notes broke the stillness, like wind chimes or a cadence of triplets piped on tiny reeds.

"Kinglets," I whispered urgently to Phil, and we were off, snowshoes skidding across the ice. I glimpsed the birds darting ahead, four grayish balls leaving the winter-bare maples for a bower of balsam fir.

We slowed and came even with them just as they spread out to feed, hopping and fluttering among the green limbs. Phil chose a random bird from the flock and began following it through the undergrowth, calling data over his shoulder in a low voice, "Side branch, side branch, branch, side branch." I took down every word. This was good stuff—time was running out on our project, and we were desperate for Golden-crowned Kinglets.

In western Maine, every January brings bitter weather, but this season marked the worst ice storm in a generation. Countless trees succumbed to the weight of their frozen crusts, toppling onto rooftops and across roads and highways throughout the region. More than thirty-five thousand power poles and pylons collapsed, leaving four million people without electricity for as long as a month. Businesses were shuttered, airports closed, the National Guard patrolled the streets, and the Rolling Stones were forced to cancel their "Bridges to Babylon" tour in Toronto, Syracuse, Montreal, and Quebec City. During all of this chaos, Phil and I hardly noticed a thing. We were staying in a remote log cabin that already lacked power, water, or road access, so the ice meant little more to us than a slippery nuisance while snowshoeing. When natural disasters strike, field camps aren't a bad place to be.

Our research was part of an excursion called Winter Ecology, a hands-on exploration of cold-weather ecosystems led by famed University of Vermont biologist Bernd Heinrich. In a career spanning everything from bumblebees and bog plants to raven behavior, Bernd has published scores of scientific papers, penned seventeen books, and made groundbreaking discoveries in how insects regulate their body temperature and how birds think. He also bakes a mean loaf of camp bread and knows how to season fried voles (a lot of salt). Every year, a dozen or so adventurous students would join Bernd at his cabin in Maine to study how plants and animals adapt to life in the cold. The curriculum relied on curiosity and chance, coming together out of whatever creatures and questions the group encountered on long rambles through the wintry forest. There was a final exam, and a paper to write, but the course felt less bounded by outcomes than by whatever the woods offered up, and the most important lesson was more basic—why do science in the first place?

"Today we entered Bernd's world of questions," I wrote in my journal at the end of our first field walk, and I overheard someone else say, "This forest is like a great big playground to him!" A consummate naturalist, Bernd used all of his senses in the woods and never restricted himself or his students to typical academic frameworks. He treated every encounter as something new, teaching science as a fundamental question mark, a raw desire to understand. "Sometimes it's better not to be a specialist," he advised us that evening as we gathered around the woodstove. Stocky and fit, with salt-and-pepper hair and bright eyes, Bernd spoke in a voice still touched by his native Germany. His approach to science retained a hint of the Old World as well, echoing the era of the great natural philosophers. "Ornithologists or entomologists—they all see one thing. But if you're a generalist, you come along with a totally new perspective, seeing things in a different way."

We spent the first week together learning a bit of everything: how to track mice, fox, weasel, and deer; where to find woodpeckers at dusk; or how to identify trees by their twigs and buds. Then Bernd turned people loose to design and implement their own research projects—small studies based not on textbooks or library searches but on daily forest walks, careful observations, and evening brainstorming sessions over cans of Budweiser. At the time, I was halfway through my master's degree and Winter Ecology felt like some kind of glorious sabbatical, a complete biological immersion in a new and fascinating landscape. I decided to study the feeding strategies of songbirds and how different species flocked together to help survive the winter. "No one has looked at that before," Bernd said thoughtfully. I interpreted that as "no one in history," but what he really meant was "no one in previous Winter Ecology courses." It didn't matter. An undergraduate named Phil Silverman volunteered to join me, and together we set off with the anticipation of true scientific pioneers.

The phrase *birds of a feather flock together* has been attributed to Plato, and in nature it's generally true. You don't find coots, pigeons, or gallinules in a gaggle of geese, and a covey of quail does not contain emus, toucanets, puffins, or wood warblers. But during winter, Black-capped Chickadees attract a crowd. In the woods of Maine, they form the nucleus of mixed-species flocks that regularly include Red-breasted Nuthatches, Brown Creepers, Hairy Woodpeckers and Downy Woodpeckers, as well as the diminutive Golden-crowned Kinglets that Phil and I were working so hard to find. The birds travel and forage together for much of the day, gathering in the morning and moving through the woods in noisy, constantly shifting groups. This habit probably developed as a way to avoid or defend against predators like owls and hawks, but we wanted to know how these birds all managed to get along so well together. Food can be scarce in winter, and it seemed counterintuitive to invite a gang of hungry rivals to dinner when the cupboard was nearly bare. How does a mixed-species flock avoid direct competition?

Our methods were simple: hike through the woods, find birds, and follow them. We focused on chickadees, nuthatches, and kinglets, carefully watching and quantifying their behaviors—how many times did each species land on the trunks, branches, or side branches of the trees? Were they avoiding conflicts by dividing up their habitat, even while foraging together? As you might imagine, counting enough songbird landings for a statistical analysis took some time, and Phil's expression became more and more pained as the days wore on. This may have stemmed in part from the cold, or the way we constantly slipped and barked our shins on the snow's icy crust. But I suspect he also began wondering why in the world he'd volunteered to chase birds with an obsessive graduate student, when most self-respecting seniors at the University of Vermont spent their last January sampling hot toddies in the better bars of Burlington. In the end, however, we had enough data to see that nuthatches foraged mostly on the trunks, chickadees dominated the main branches, and kinglets spent their time flitting about in side branches. It was a neat example of what ecologists call "niche partitioning," using subtle variations in behavior to divide a resource among potential competitors.

While the niche study made for good conversation at that year's Maine Bird Conference, Winter Ecology left me with an even stronger memory about feathers. One night at the tail end of the ice storm, the skies cleared, and the temperature plunged to seventeen degrees below zero (–27° Celsius). That's cold enough so that a Budweiser, spilled in the snow, will freeze solid before all the beer can drain from the can. I know this because I dropped one on the walk between Bernd's cabin and my tent. Still, it was a beautiful night. The maples and pines bowed earthward, their heavily iced branches glowing in the moonlight like blown glass. Readying myself for sleep, I couldn't help thinking of the hapless "chechaquo" in Jack London's Yukon tale, "To Build a Fire." But where he had hoped to ward off the frost with "mittens, ear-flaps, warm moccasins, and thick socks," I had a plush goose down sleeping bag to burrow into. (I'd borrowed it from a friend who was sleeping indoors beside the woodstove.) Lying there in warmth and comfort, it occurred to me that somewhere nearby, the tiny songbirds that Phil and I had been chasing all afternoon were doing exactly the same thing.

The thought of any creature surviving outdoors in such frigid temperatures is impressive, but the Golden-crowned Kinglet does it with the smallest body mass of any bird in the north woods. In ecology, Bergmann's Rule states that body size generally increases with latitude—larger species flourish in colder climates because bulky things maintain their temperature more efficiently. Put a big pot of stew out in a snowstorm, and it will stay hot a lot longer than a grilled cheese sandwich or a fried egg. A Golden-crowned Kinglet, however, weighs in at just over five grams, about the same as a nickel or a teaspoon of salt. That's less than half the size of the chickadees and nuthatches it flocks with, and where those birds huddle for warmth in empty nest holes at night, kinglets appear to make do in the open air. Phil and I followed them several times at dusk and once tracked a pair to the crown of a young balsam fir. As darkness fell I tried climbing to their roost but couldn't force my way through the dense branches. Still, the tree was certainly too small to harbor any kind of cozy nest cavity; at best their shelter for the night consisted of a fir branch covered by a thin layer of snow. Shivering is one strategy to generate body heat, and some species can slow their entire metabolism at night, entering a kind of low-temperature torpor to pass the time until sunrise. But clearly only one thing keeps kinglets and countless other birds from freezing solid in the coldest climates: the incredible insulative quality of feathers.

Heat transfer occurs in three ways: by wavelengths through empty air (radiation), by the movement of air (convection), and directly through an object to adjacent surfaces (conduction). A campfire warms you by radiation, the heater in a car works by convection, and when you burn your tongue on a slice of hot pizza, that's conduction. Insulation slows down all three of these processes by trapping air and using it as a barrier—the more static air a material can hold, the better it is at keeping the hot side hot and the cold side cold. Sleeping-bag manufacturers call this quality loft; in the construction industry it's called R-value. A complex, fluffy substance with a lot of interior surface area

A Golden-crowned Kinglet.

works best at capturing pockets of air molecules and making them stay put. This is why ski parkas are puffy and why fiberglass house insulation looks so much like cotton candy. It's also why a kinglet, though small, looks a lot bigger than a teaspoon of salt. With their intricate air-trapping microstructure, down feathers are the most naturally insulative material on earth, and birds have the ability to fluff them up manually, essentially adjusting their R-value at will.

Rick Prum's theory suggests that down feathers evolved quite early, and their importance to birds is revealed by simple math. Of the two to four thousand feathers that cover the average songbird (or the twenty-five thousand on a Tundra Swan), the vast majority include downy basal barbs or feature downy appendages called afterfeathers, and many are entirely downy in structure. Feathers adapted for flight, in contrast, number only a few dozen. When safely tucked beneath a bird's weatherproof contour feathers, down traps countless pockets of warm, dry air near the skin and makes life in the cold possible. Kinglets in a Maine ice storm offer an extreme example, but all birds need to survive unpredictable climatic events and large fluctuations in temperature. The quantity and quality of their down corresponds directly to their environment and lifestyle, and they manipulate the feathers to trap or release heat depending on the weather, season, and time of day.

In the years after I attended Winter Ecology, Bernd became increasingly fascinated with Golden-crowned Kinglets and their ability to withstand the cold. The little birds enjoyed a starring role in his book *Winter World*, where he described studying their stomach contents (moth caterpillars), calculating their caloric burn rate (thirteen calories per minute), and searching tirelessly for their roost sites. When he looked at a kinglet's feathers, he found that the vast majority of its plumage was devoted to insulation, accounting for a full 7 percent of its body weight. The difference between the outdoor air temperature and the cozy space inside a kinglet's feather coat could be as large as an astonishing 140 degrees Fahrenheit (78 degrees Celsius). Deprived of its feathers on a frigid night like the one I passed in my comfy sleeping bag, a perched kinglet would freeze solid in a matter of minutes, almost as quickly as a spilled can of beer.

Ample evidence for the virtues of down can be found in the distribution of birdlife on planet Earth. More than 300 species visit the tundra and rocky islets above the Arctic Circle, and 240 live on the frigid high plains of the Gobi Desert. Emperor Penguins incubate their eggs standing outdoors unprotected in the middle of the Antarctic winter, and the Bar-headed Goose regularly flies above thirty thousand feet in its annual migration over the Himalayas. At that altitude, air temperatures and wind chill often combine to dip below –80 degrees Fahrenheit (–62 degrees Celsius). Like kinglets, these birds all rely on feathers to insulate their bodies against the elements. But the usefulness of down does not stop at the skin of a bird—an equally powerful testimonial lies in the countless ways that other creatures have co-opted them for insulation.

The view from my desk in the Raccoon Shack takes in a corner of our orchard, a fence line, the edge of a willow grove, and a bit of old pasture growing up with alder. One spring afternoon I watched a female Yellow-rumped Warbler emerge from the willows again and again to land on the fence, directly above a hollow where the chickens had taken dust baths. Each time she would look around warily, then dive to the ground and hop up again with a feather firmly clamped in her bill. I didn't need to find her nest to know it was nearly complete, its cup neatly lined with the shed plumes of our laying hens. For just as a kinglet knows to fluff up its down on a cold night, so have birds everywhere realized that *extra* feathers will help keep their hatchlings warm.

A full quarter of North American bird species use feathers in their nests, including nearly all ducks, wrens, and swallows, as well as many warblers and finches. Canada Geese and Great Horned Owls pluck down from their own breasts, but most birds follow the Yellow-rumped Warbler's strategy, scavenging plumes cast off by other species. Kinglets themselves are accomplished feather builders, often gathering their nesting material in the wake of a sparrowhawk or falcon attack on a nearby flock. (Raptors typically pluck their avian prey immediately after the kill, leaving a neat pile of feathers ripe for the taking.) In a study of the Goldcrest, a European kinglet, three nests were painstakingly deconstructed and found to include an average of 2,611 individual plumes. Young birds brought up in feathered nests benefit from the added warmth, growing larger and fledging sooner than those raised without. The plume layer also helps protect them from parasites. In fact, a whole host of creatures take advantage of borrowed feathers. Deer mice and white-footed mice stuff their burrows with them, and bumblebees will reuse that same material after the rodents move on. In desert regions, pack rats accumulate piles of feathers in their sprawling middens, where the combination of a dry climate and crystallized rat pee can preserve them for thousands of years. One species, however, stands out above all others for its creative co-option of feather insulation. From sleeping bags to comforters, pillows to parkas, quilted hats to booties and doggie beds, only *Homo sapiens* has turned down and feathers into a multibillion-dollar global enterprise.

To gain a better understanding of the down industry, I took the obvious first step: I read the label on my pillow. It showed a handsome white goose on a blue and green background with the words *Pacific Coast Feather Company* emblazoned on a gold banner. Surprisingly, the address read Seattle, Washington—only a ferry ride and a two-hour drive away. I dialed the phone number and explained my situation to a friendly receptionist. She patched me through to the right people, and within moments I had an appointment for a guided tour of the largest down and feather factory in the country.

Run by the same family for generations, Pacific Coast Feather makes and sells millions of pillows and comforters every year—a sizable portion of the North American market. They operate thirteen manufacturing plants in ten states and Canada, but every feather in every one of their products comes through one facility located at the end of a quiet side street north of Seattle. I arrived on the first day of autumn, crossing the parking lot through a carpet of poplar leaves that drifted and rattled in the wind. Then I noticed something drifting with them—a scattering of white feathers and down that grew thicker the closer I got to the factory. Working with materials as light as air made spillage inevitable, and it was no wonder they'd moved their factory to a rural area. Everything used to be processed at their downtown headquarters, where the occasional broken duct had been known to shower central Seattle with an unseasonable snowstorm of fluff and plumes.

"We process about two hundred thousand pounds of feathers here every month," my host told me. His name was Travis Stier, an affable, solid-looking fellow in his late thirties. He had a toothy smile that often began in midsentence, so that by the end of a story he was laughing out loud. As Pacific Coast Feather's chief buyer, he regularly negotiated deals that accounted for more than 5 percent of the world's annual production of goose and duck feathers. They came from China, Thailand, Vietnam, France, Hungary, Poland—dozens of countries where waterfowl feature prominently in the local cuisine. Travis had been in the feather trade for sixteen years, but that still made him one of the younger buyers on the circuit. "It's a bit of an old boys' network," he admitted. "You have to have an instinct for the business."

Unlike most industries, where the demand for finished products determines the supply, the feather and down business has a supply chain completely unrelated to its market. Demand comes from people buying pillows, comforters, sleeping bags, duvets, and other high-end consumer goods. Supply, on the other hand, is determined almost entirely by the consumption of goose and duck meat, primarily in rural China and Southeast Asia. When people eat a lot of birds, there will be plenty of feathers to buy. When consumption falters, tastes change, or birds have a bad year, then the feather supply comes up short. And the timing matters, too. If farmers slaughter their fowl too soon, before the birds add extra down during the fall molt, then even a good supply of meat birds won't yield enough feathers. Add to this the unpredictability of disease outbreaks like bird flu, and feathers become one of the most volatile of all commodities. I asked if anyone ever tried to corner the feather market, and he laughed. "All the time!" Just the previous year, Travis had watched prices triple in a matter of months as Chinese speculators bought up every bit of fluff they could find—just one more hitch on the long trip from bird to pillow.

"Every village in China has a guy on a bicycle riding around from farm to farm, buying feathers," Travis explained. He described the surprisingly spacious farmhouses and how a family could amass quite a store of feathers right in their living room before deciding to sell. Ducks are a staple and make up the vast majority of down supplies, but geese are popular menu items, too, and their feathers generally bring a higher price. Meat was the primary reason to raise any bird, but the feathers were an important added value that people paid close attention to. Hard negotiations began at the level of rural farmer and buyer and repeated themselves at every step of the chain. Once enough bicycle loads had been amassed to fill a truck, the local buyer sold his stock to a regional factory, where some degree of washing and sorting took place. The small factories sold in turn to several large processors, who also bought the masses of feathers coming out of China's more industrial-scale poultry farms. Travis struck his deals with the large factories, building on relationships that his predecessors at Pacific Coast Feather had established in 1972, immediately after Nixon opened up a trade window in the "Bamboo Curtain."

"I like to make the feather business work better," Travis said at one point, and I glimpsed a tenacity beneath his outer calm—something that probably made him an excellent negotiator. "It's so easy to cheat—adding sticks, rocks, or sand to the product at some point along the way. Everything is sold by weight!" He told me that his first job at Pacific Coast Feather had been ferreting out the bad seeds in their supply chain, eliminating anyone whose product didn't meet his standards. "Everyone is on board now," he summed up with a satisfied nod. "It's really not a problem anymore."

Before heading down to the factory floor we stopped by a small laboratory where quality checks were under way on a recent shipment from Taiwan. A burly, familiar-looking man stopped what he was doing and cheerfully showed us around. ("My brother, Jon," Travis explained later. In spite of its large market share, the company still maintains a decidedly homey feel.) Filing boxes brimming with feathers sat waiting to be inspected. The highest-quality down quivered like something liquid, and bits of it floated through the air all around us. I plunged my hand into a typical pillow mixture—mostly contour feathers from the neck and body, with a bit of down blended in. It felt soft, still, and immediately warm. "Duck," Travis said, and then frowned at the pile of identical white and blackish plumes, "but I see a few chicken feathers in there."

As we entered the main hangar of the factory, the dusting of feathers I'd noticed in the parking lot swelled to a veritable tide that swirled around our feet with every step. There was a man raking great tufts from a stack of six-foot bales toward a loading bay where two pipes vacuumed them up and spat them into the production line. Behind him, a row of huge green machines whirred and chugged, venting steam toward the ceiling, where more pipes and ducts crisscrossed in a crazy network. I saw wooden-framed dryers and chutes, hoses, wires, rows of mysterious doors, and in the distance a pair of three-story towers lined floor to ceiling with windows. All in all, it looked like a place where Willy Wonka would have felt right at home.

When I asked Travis if I could take pictures, his laugh rang out over the din. "All you want," he assured me. "We don't have any secrets—feather technology hasn't changed in a hundred years!" Many of the machines did have an old-fashioned look, like fine cabinetry with their framed windows and varnished wooden hatches. A hundred years ago, however, there would have been a lot more guys moving feathers around with rakes and pitchforks. Modern automation pumped everything through with piped air and required only six workers to keep the line humming. I snapped a few photographs as we walked through, but nothing could capture the commotion of the place or its distinctive smell—an animal odor, not rank but faintly rich, like a room where a pot of soup had been simmering earlier in the day.

"Three things happen in a feather factory," Travis explained, "Washing, separating, and mixing." He showed me the large drums where feathers were steam cleaned and rinsed at least six times to remove any trace of dirt or dust. (Contrary to popular opinion, research now shows that properly cleaned down is not allergenic—it's the dust and associated mites that people react to.) Once dry, the raw down and feather gets blown into separation chambers, the two huge towers at the far end of the building. A fresh batch dropped in just as we approached, and every window held its own whirling vortex of white plumes. The effect was beautiful and a bit hypnotic, like staring into the spin and drift of dozens of newly formed galaxies. Since the best-quality down rises highest in the wind, the towers worked by a simple series of baffles and breezes that neatly separated the raw mixture into as many as seven different grades. Feathers too large to use dropped to the bottom, where they were ground into waste.

I understood how the machines worked, but their appearance still mystified me. The proliferation of windows was wonderful to look at but seemed like overkill—surely, a few well-placed openings would have been enough to check on what was happening inside. Only later did I realize that their design fitted a larger pattern: people who work with feathers like to see them. Whether a person is tying flies, sewing hats and costumes, studying aerodynamics, or working in a factory, feathers invite a fascination that transcends the toil of their daily grind. We stood there a while watching the feathers drift and settle. "I saw a separator in China once that was made entirely of glass," Travis remarked.

A three-story sorting machine at the Pacific Coast Feather Company factory.

I asked if there wasn't some kind of framework holding it up. "No," he said, still staring up at the feathers. "The whole thing was glass."

At the end of the tour we shook hands, and I took a picture of Travis in front of stacks of the finished product, huge bales of sorted feathers ready to be shipped out to plants across the country. Tufts of down clung to the outside of the bales, and I could see bits of it drifting out the loading-bay door and into the breeze.

That impossible lightness was the key to its insulative qualities—an unsurpassed ability to trap air. Walking back across the parking lot, I stooped and fished a few plumes out from among the orange autumn leaves. I found a perfect down feather, its sprawling barbs waving like the tentacles of some fantastically delicate anemone. It seemed a shame that its long journey from the breast of a Chinese farm duck should end in a parking lot. I found myself tempted to take it back inside, so it too could help keep someone as warm as kinglets.

Certainly, nothing else can do the job as well. An Eskimo's caribou furs provide plenty of warmth, but a full storm suit with matching mukluks and mittens weighs in excess of eighteen pounds. Synthetics are lighter than animal pelts, but a mountain climber would still have to wear eleven pairs of polypropylene long johns to achieve the same heat retention as one down-filled expedition jacket. Top-quality goose down is more than twice as efficient by weight as Thinsulate, Polarguard, Primaloft, or other synthetics and durable enough to withstand years of heavy use. Its secret lies in its structure, the way each feather branches and branches again into an intricate web of air-and heat-trapping barbs and barbules. Feathers grow that way naturally, but manufacturing such finely branched filaments is extremely difficult. Synthetic fibers can be hollow, coiled, kinked, or woven into complex fabrics, but they remain fundamentally single-stranded, an insulative disadvantage that no amount of engineering has so far been able to overcome.

In spite of this, down makes up only a small portion of the market for insulated products, and there's little chance that feathers will drive the polypropylene people out of business. This is due in part to supply limitations—the world would have to eat a lot more stewed goose to fill every last sleeping bag, doggie bed, and winter coat. But something more fundamental prevents feathers from total domination of the industry: water. When down gets wet, it soaks up the moisture, collapsing those tiny pockets of warm air into a sodden mess that loses much of its insulative value. It also requires a long time at low humidity to dry out and refluff, which can be difficult to accomplish when huddled inside a tent in a rainstorm. Birds avoid this problem by keeping their down safely tucked away under layers of waterproof contour feathers, but for outdoor clothing and camping gear, the system is imperfect. Synthetic fibers, on the other hand, are generally petroleum based and hydrophobic, repelling water and retaining their insulative properties even when fully submerged. Current research in the insulation industry centers on combining the best of both worlds, finding a way to build artificial featherlike structures from water-resistant materials. It's a kind of Holy Grail quest, and no one in the down industry seems terribly worried. "The synthetics have gotten a whole lot better," Travis Stier told me, "but they're still heavier than down. And they don't breathe the same way."

On the final night of my Winter Ecology experience, I hiked into Bernd Heinrich's woods to admire the ice and moonlight one last time. The weather had warmed up considerably, but temperatures still hovered around 0 degrees Fahrenheit (–18 degrees Celsius), and I was dressed for maximum warmth: wool socks, lined Snow-Master boots, long johns, a pair of Ranger Whipcord wool trousers, scarf, hat, gloves, and three shirts topped by a thick yak-wool sweater. (The lack of feathers in my graduate-student wardrobe can be explained by another drawback to goose down: its price!) The crisp air was utterly still, interrupted only by the sound of my breathing and the occasional crack of a branch giving way under its burden of ice. Whenever I started to get cold I walked faster or ran a few steps until the exercise warmed me up again, and it occurred to me that the real challenge for kinglets was not simply keeping warm but *regulating* their temperature. Winter days in Maine might fluctuate by twenty or thirty degrees, and the summer months would bring heat waves more than one hundred degrees higher. In desert climates, birds can face those kinds of extremes in a single twenty-four-hour period. If feathers had evolved so well for insulation against the cold, how did they function when things heated up?

CHAPTER SIX

Staying Cool

The hyperthermic bird must increase ventilation rates over evaporative surfaces in the respiratory system and buccopharynx for cooling purposes and yet avoid a severe blood hypocapnia and alkalosis.

—Mike Gleeson,
Analysis of Respiratory Pattern During Panting in Fowl (1985)

The eagle attacked again, swooping down and slamming into the water with a great splash exactly where the cormorant had been resting a split second earlier. Waves spread outward in a ring as the eagle heaved itself skyward, wings pumping hard, plumage dripping, talons empty. I scanned the surface of the pond and saw the cormorant pop up near the shoreline with its head held low. It scrambled out onto a dock as the eagle resumed its perch high up in the winter-bare maple. There they sat, like two boxers in their corners after a particularly hard round, glassy-eyed with tension and fatigue.

Bald Eagles prefer to scavenge or steal their meals whenever possible. If forced to hunt, they target weak, spawning fish; helpless nestlings; or small mammals. They will go for other birds in a pinch, but chasing down this savvy Double-crested Cormorant proved no easy task. I watched the battle drag on for nearly an hour, the eagle apparently trying to harass its quarry to death, dive after dive. For its part, the cormorant played a risky strategy, submerging only an instant ahead of the eagle's claws as if daring the huge raptor to join it underwater. In the end, both birds looked equally exhausted, and I'll never forget how their bills hung wide open, tongues protruding. It was the first time I'd ever seen a bird pant. The eagle finally gave up, but only after one plunge too many, when it couldn't flap itself airborne again and had to swim pathetically to shore, paddling with its long wings.

This incident played out on a small lake where I'd gotten a great deal renting a summer cabin in the middle of winter. The place was freezing and I was short on firewood, so at the time I was preoccupied with discomfort and could only think about how cold those poor wet birds must have been. Now, thinking about feathers, I wonder how they didn't expire from heat stroke.

Muscle activity burns calories, releasing that energy as both movement and heat. The more strenuous the motion (such as flying and diving after your dinner, or swimming for your life), the greater the potential increase in body temperature. All animals function this way, which is why long-distance runners prefer shorts and a T-shirt to yak wool and mukluks. Mammalian athletes also help cool themselves by sweating, but birds lack that ability. What's more, birds already operate on the hot side, maintaining a basic metabolic rate considerably higher than that of mammals. The Golden-crowned Kinglet, for instance, keeps its body temperature at a steady 111 degrees Fahrenheit (44 degrees

Celsius) year-round. That's within a few degrees of the point at which proteins in living cells begin to break down faster than the body can replace them, a situation that quickly leads to disorientation, loss of consciousness, and death. So birds can't afford to heat up, which makes their ability to withstand the warmth of vigorous exercise and life in hot climates just as amazing as their adaptations to life in the cold. What's more, they manage to do it while fully encased in feathers, nature's finest insulation. It would be like running marathons dressed in a sleeping bag.

The problem of keeping a feathered bird cool can be broken into two distinct parts: coping with internal sources of heat (e.g., muscles working during exercise) and coping with external sources (e.g., sunshine). I decided to tackle the second question first—it seemed like something I could answer with a couple of thermometers and a dead woodpecker.

If you ever decide to write a book about feathers, you'll find that people begin saving them for you—individual plumes, disembodied wings, and sometimes whole bird carcasses that they've found beneath their window panes, wrested from their cat, or picked up by the side of the road. This activity is technically illegal in the United States—most wild birds, as well as their nests, eggs, and feathers, are protected from any harassment or handling under the Migratory Bird Act. But still the specimens trickled in. The woodpecker arrived on our porch neatly wrapped in a trash bag and still quite fresh when I tucked it into the freezer for safekeeping. It was a Northern Flicker, a species that ranges from mountainsides above the Arctic Circle to the tropical forests of Central America and the Caribbean. It even survives the heat of Death Valley, California, a sunken desert basin where summertime temperatures can reach 134 degrees Fahrenheit (57 degrees Celsius). If any plumage was designed to help a bird stay cool as well as keep warm, this had to be it.

Northern Flickers, by John James Audubon.

I thawed my flicker on a late summer day and let it warm up on the shady porch of the Raccoon Shack. Looking at the bird reminded me of the sheer wonder inherent in even the commonest things. With its ubiquity and tame habits, the Northern Flicker rarely merits more than a glance from nature watchers, but up close the hues of its plumage shone as if lit from within. It was a male, with a crimson streak of tiny feathers that raked back from its bill into the ashy wash of its cheeks and throat. The top of its head graded from bright chestnut above the eyes to an olive nape, which then faded into the rich brown and black brindle descending its back. I lifted its folded wings to reveal a rosy sunrise blush beneath every feather and the white patch on the rump that was so distinctive in flight. Even the breast feathers were decorated, each creamy plume splashed with a dollop of black, lustrous and perfectly centered on the tip of the vane. An artist with strong imagination and an unlimited palette could hardly have devised such a vision.

I took the dead bird's temperature in the shade of the porch, placing thermometers both inside and outside its beautiful feather coat. In both cases the gauge registered a comfortable 74 degrees Fahrenheit (23 degrees Celsius), confirming that the specimen was fully thawed and acclimated. Then I brought it out into the sunlight, propping it in the mowed grass in the upright position I often saw flickers use when ground feeding. I could feel the sun's radiant energy warming my skin, and the flicker's dark feathers quickly grew hot to the touch. I put one thermometer against the outside of the back feathers and tucked the other beneath them, up under the wings and into the contour feathers and semiplumes close to the skin. After a few minutes, the temperature at the surface of the back feathers surged to 102 degrees Fahrenheit (39 degrees Celsius), but inside it reached only 87 degrees Fahrenheit (31 degrees Celsius). A half hour later, the temperature inside the feathers had increased by only two more degrees—warmer than the shady porch but considerably cooler than the bird's dark exterior, less than an inch away.

By blocking the sun's radiant energy and slowing down conduction with a complex baffle of feathers, the flicker's plumage achieved a thermal reduction of between 13 and 15 degrees. For the sake of comparison, I leaned a large quarter-inch ceramic tile up next to the bird. I got it from beneath the Raccoon Shack's woodstove, where I'd been using it to help protect the floor from excessive heat. Its surface was a lighter shade than the flicker's back and reached only 97 degrees Fahrenheit (36 degrees Celsius), but conduction quickly warmed the whole tile, and it soon measured 92 degrees Fahrenheit (33 degrees Celsius) below, only 5 degrees less than the top. In this simple experiment, the plumage of an inert Northern Flicker performed two to three times better than an insulating ceramic tile. For live birds, with their ability to change body position, as well as lift, fluff, and adjust their feathers, a whole host of strategies exist to make feathers an even more efficient heat shield.

For decades, some experts promoted the "heat shield" idea as the genesis of feather evolution. It started with herpetologists, who noted that various iguana-like lizards had long, raised scales whose tiny shadows helped them avoid overheating in hot climates. The theory went that these tiny heat shields continued elongating in the ancestors to birds, fraying from the edges, becoming flexible, and finally losing their scalelike qualities entirely and evolving into all the diverse forms of modern feathers. This turned out to be nonsense, but that's not to say that lizards and birds don't share several basic temperature-regulating behaviors. As cold-blooded creatures, lizards rely on the sun's rays for warmth and alter their body position meticulously throughout the day to keep things within their comfort zone. Mornings may find them basking broadside for maximum heat, but as the day warms up, they will face parallel to the sun to minimize their exposure. When things get really hot, a lizard rises up on its feet to increase convective cooling and may retreat entirely into a shady crevice to pass the heat of the day.

Birds do all of these things, though seeking shade is usually the first rather than the last line of defense. It's no coincidence that bird-watchers time their efforts for early morning and just before dusk, since nearly all species concentrate their activities in those coolest parts of the day. Had my Northern Flicker been among the living, there's little doubt it would have sought a shady perch rather than sit still in the grass while its back feathers grew hotter and hotter. But some situations force birds to remain in full sunlight, and then they follow the lizard's excellent

A girdle-tailed lizard.

example of strategic body positioning. Griffon Vultures in Africa and Asia often wait for hours at exposed perches near kill sites, but simple changes in orientation and posture can limit their sun exposure while at the same time quadrupling the amount of bare skin revealed to any breeze, greatly increasing convective cooling. (California Condors and other New World vultures follow suit, but also defecate on their own feet to gain the added cooling effects offered by evaporating excrement.) Herring Gulls and many other seabirds nest in the open on rocky islets and cannot abandon their eggs, even on the hottest days. Instead, they rotate slowly on their nests like a field of feathered sunflowers, keeping their backs carefully parallel to the strong beams of sunlight. One famous study identified six distinct cooling strategies employed by nesting Sooty Terns, from panting to ruffling their back feathers to rising up and exposing their bare legs.

Heat-avoiding behaviors and the basic protections offered by a feather coat can help mitigate the effects of a sunny day, but birds face an even greater challenge from the heat generated within. It's an unavoidable by-product of the sheer physical exertions of their daily lives. No vertebrates on earth maintain a higher metabolism and generate more heat than birds. Flight muscles alone make up as much as 35 percent of a bird's body weight and give off up to

A Sooty Tern on its nest uses at least six techniques to stay cool: (1) exposing the wing bend apteria, (2) panting, (3) ruffling crown feathers, (4) ruffling back feathers, (5) evaporation from wetted abdomen, and (6) shunting blood to exposed legs.

90 percent of the energy they use as excess heat. When a bird takes flight, it suddenly finds itself producing seven, ten, or even twenty times the body heat it had while perched. (The large leg muscles of running birds tell a similar story, with rheas and ostriches showing huge metabolic increases at top speeds.) Though feathers can be lifted or ruffled to let out a bit of warmth, birds clearly must have other physical adaptations that help them burn so hot yet stay cool.

Innovation in nature often occurs at stress points, places where competing adaptive pressures create an evolutionary dilemma. Through flight and fast running, birds generate tremendous body heat while living inside a coat of the world's finest insulation. If we accept the growing consensus that feathers evolved in theropod dinosaurs and that the downy, insulative types came before flight feathers, then it's reasonable to assume that powered flight and specialized cooling mechanisms developed in tandem. As early fliers expanded their flapping capabilities and produced more heat, their already insulated bodies evolved ways to get rid of it. And if we accept John Ostrom's theory that dinosaurs were active, warm-blooded creatures, then the basics of avian cooling must have already been in place in theropods, whose fleet-footed lifestyle generated plenty of muscle heat in its own right. Both of these premises appear to be true, and the result is a complex system of feather manipulation, controlled blood flow, and evaporative cooling that allows most birds to dispel far more heat than they produce, even while flying on a warm day.

A bird's cooling strategies work in close cooperation, but they fall into three distinct categories. The first has to do with the position of the feathers themselves. On the day I dismantled my Winter Wren down at the Raccoon Shack, one of the first things I noticed was how the feathers were distributed. Though arranged to cover the whole body, their bases grew from discrete tracts separated by large areas of bare skin called *apteria.* When birds heat up, they can raise and shift their feathers to reveal these bare patches, allowing air and wind to whisk away body heat through convection. And while birds may lack sweat glands, they do have the ability to release water directly through the skin of these patches, greatly increasing the efficiency of heat loss. Among the very few birds to lack distinct apteria are penguins. Their feathers are distributed evenly and thickly across their bodies, but they live in a frigid environment with plenty of other options for dispersing heat.

When human bodies warm up, small blood vessels near the surface of the skin widen automatically, allowing more arterial flow to transfer heat to the outside air. That's why people's faces look flushed when they're jogging up a hill or sitting in a sauna, and why excessively red skin is a classic symptom of heat stroke. Birds do the same thing. But with more heat to disperse and few

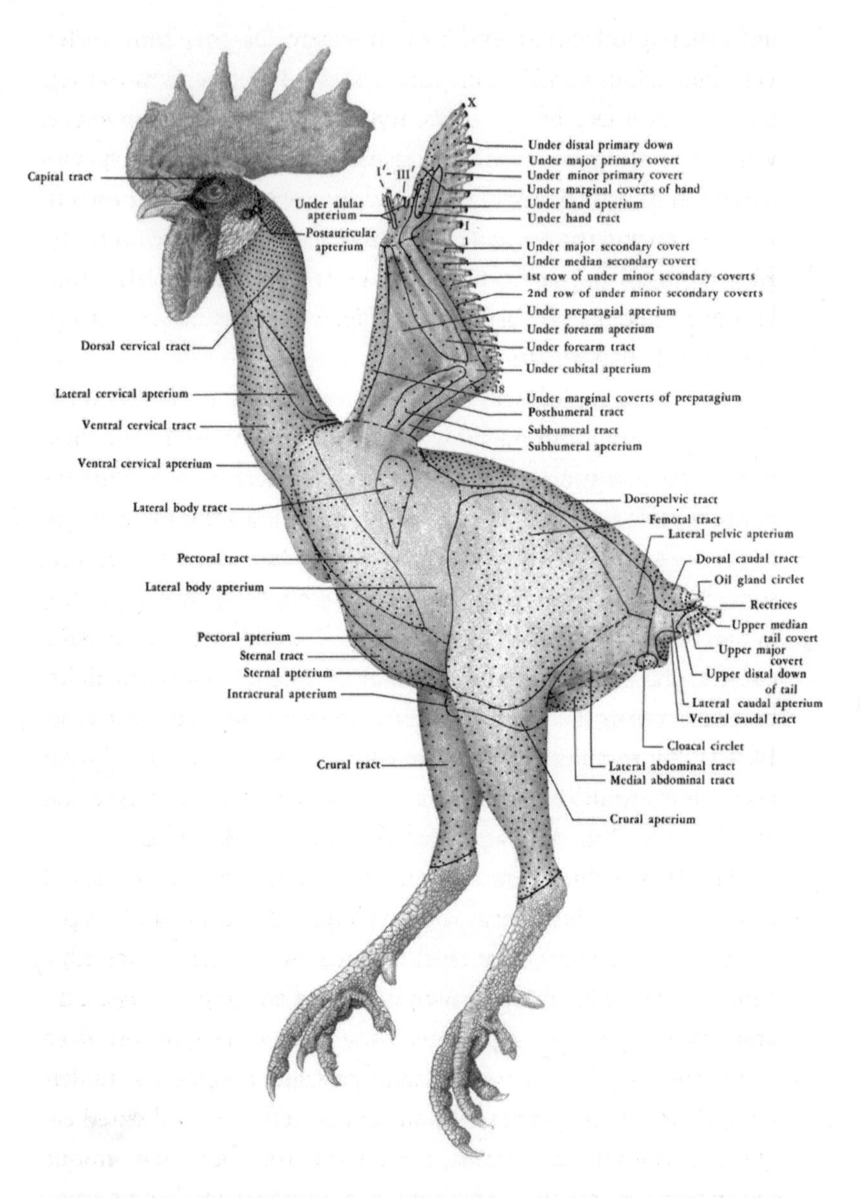

The intricate pattern of feather tracts (*pterylae*) and bare skin (*apteria*) on a Single-comb Common Leghorn rooster.

unfeathered surfaces to send it to, they move blood around with a vengeance. Gulls and herons can increase the blood flow to their bare legs and feet by up to twentyfold, quickly radiating excess warmth to the surrounding air or water. Scores of other species dangle their legs to dissipate heat during flight, a habit known to increase during the warmth of midday for tropical birds like the Blue-throated Bee-eater. In experimental trials, artificially insulating the legs of a flying Common Pigeon quickly caused it to go dangerously hyperthermic.

Heat-stressed birds will also shunt more blood to their apteria, to the bare skin around their eyes, and to wattles if they have them. Blood vessels in the bill release body heat as well. This explains why so many species tuck their beaks into their feathers at night or during cold weather, but it may also account for why some birds in hot climates developed such grossly oversized bills. When researchers photographed Toco Toucans with an infrared camera, the birds' clownish beaks lit up like incandescent lightbulbs whenever the temperature rose beyond their comfort zone. Blood flow to the bill accounted for 30 to 60 percent of their body's total heat loss, making toucan bills one of the largest and most efficient "thermal windows" in the animal kingdom.

The final strategy in avian cooling relies on the physics of evaporation. It takes energy to turn a liquid into a gas, and evaporating water droplets draw part of that power as heat from adjacent surfaces. This is the principle behind panting in dogs, cats, and other mammals—rapid breathing evaporates moisture from inside the mouth, throat, and nasal passages, cooling the underlying tissues in the process. As any exhausted and overheated eagle or cormorant can attest, birds pant, too, but their unique respiratory system takes evaporative cooling to a whole new level.

When birds inhale, the air takes a detour. It doesn't simply fill their lungs and wait patiently to be breathed out again. It starts a

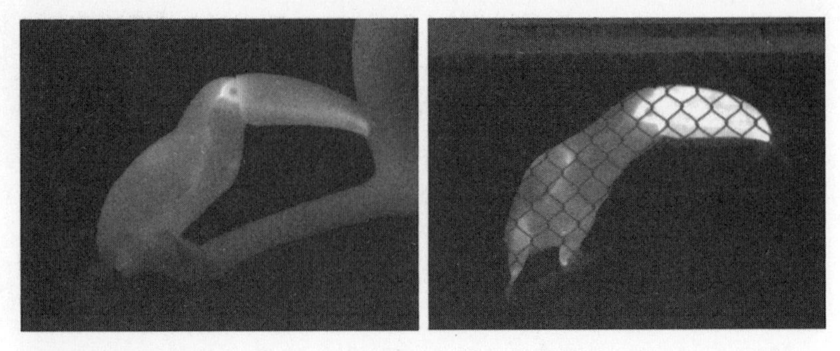

Thermal images of a Toco Toucan in cool conditions (*left*) and warm (*right*), when its bill begins radiating excess body heat.

four-step process that keeps air flowing continuously throughout their body cavity, and even into their bones. Unlike mammals and most other vertebrates, who breathe with a straightforward in-and-out motion, birds have evolved a complex system of nine or more air sacs to supplement their lungs. These sacs surround the internal organs and extend inside their leg and wing bones. (Fossil evidence of similar sacs in theropods provides another link in the bird-dinosaur story and confirms Ostrom's idea that hot-blooded avian ancestors already had need of a good cooling system.) A bird taking a breath brings air into the first network of sacs; breathing out again then pushes that air into the lungs. The next inhalation moves things on to the forward sacs, and only after the second exhalation does that original beakful of air go back out through the bird's mouth or nostrils. This system increases the efficiency of the lungs, but it also dramatically expands the surface area available for internal evaporation. It brings air into and around the very muscles generating the heat and allows for rapid and widespread cooling with every breath. And when a panting bird can take hundreds of breaths per minute, this makes for a highly efficient system. Wind-tunnel experiments show that evaporative cooling alone can dispel a quarter of the heat produced by a hovering hummingbird and nearly half that generated by flying parakeets. Used in conjunction with leg dangling and directing blood flow to thinly feathered thermal windows, the mere act of breathing helps keep an active bird from overheating.

To fully appreciate how feathers helped shape the avian cooling system, one need look no further than bats, the only other living vertebrates capable of powered flight. Like birds, bats experience a huge increase in heat production whenever they launch themselves airborne and begin to flap. They too have large flight muscles whose metabolic demands and outputs match those of birds almost exactly. But where birds transformed their respiratory tract into an elaborate and unique inner evaporation system, bats make do with the same simple lungs and cooling techniques used by house mice, stoats, or any other mammals. Without feathers, bat bodies retain a suite of much simpler cooling options. Their huge, naked wings are the most obvious—natural thermal windows that can be pumped full of hot blood and fanned in the breeze. But even that is only necessary in extreme situations. Studies of captive bats found them perfectly capable of maintaining a comfortable body temperature on short or moderate flights, flooding their wings with blood only when forced to stay aloft for a half hour in a heated chamber. A bat's fur is only moderately insulative, and they rely more on roosting and huddling in caves and crevices to survive cold snaps. In flight, however, this thin coat allows them to radiate heat directly from their bodies. Thermal images tell the story perfectly. Bats bodies glow all over like bright coals, while birds radiate their heat from only a few hot spots: bill, legs, and a lightly feathered area under each wing.

On the left, Brazilian free-tailed bats (*Tadarida brasiliensis*) radiate heat from across their bodies while the Barn Owl (*right*) loses heat only from its bill, legs, and the apteria beneath its wings.

Overcoming the thermal challenge of insulated flight has helped give birds an incredibly wide ecological range. Individual species like the Northern Flicker can thrive from the tropics to the Arctic, while birds in general occupy every continent and roam the open oceans, surviving virtually any climate the planet has to offer. Bats stay cool with comparative ease, but that ability comes at a cost. Though among the most numerous of mammals, bats are restricted to moderate conditions and habitats—remaining nocturnal to avoid the heat and migrating or hibernating en masse to avoid the cold. To thrive in colder climates, bats would need better insulation, but that in turn would require a better way to stay cool in flight.

The close relationship between avian cooling and flight comes as no surprise. Sooner or later, nearly any discussion about birds turns to this most basic of their traits, a habit so closely intertwined with feathers that our story must turn to it directly. In the next section we will explore just how flight may have evolved, and what role feathers played in the process.

Flight

There is an art . . . or rather, a knack, to flying. The knack lies in learning how to throw yourself at the ground and miss.

—Douglas Adams, *Life, the Universe, and Everything* (1982)

CHAPTER SEVEN

Ground-Up or Tree-Down?

GINGER: *We haven't even lifted off. Why?*
MAC: *Throost! . . . Wot we're miss'n' is throost.*
ROCKY: *I didn't get a word of that. . . . I swear she ain't using real words.*
GINGER: *She said we need more thrust.*
ROCKY: *Thrust!?! Of course we need thrust!*

—Chickens discussing aerodynamics, from *Chicken Run* (2000)

Before Eliza and I had a child, we had chickens. And to a certain degree, acquiring them resembled the process of bringing a baby into the world. First came a long gestation period of planning, reading, discussing, fencing, and multiple trips to the Web site "Findmychicken.com." This was followed by the challenges of birth, as the two of us science types struggled with the practicality of hammer and nail to build a functional coop. After a trip to the county fair and a fierce bidding war at the 4-H livestock auction,

Trouser, Fatty, and White One finally arrived. We got them as pullets, young hens a few months shy of laying eggs and still very much learning the art and Zen of chickenhood. They didn't know how to scratch, they couldn't cluck properly, and they seemed entirely unsure of whether they should be flying like the sparrows that flitted nimbly in and out of their coop, making off with beaks full of grain.

When we first got them home, the birds spent much of their time dashing madly about the orchard, flapping their stubby wings and even lifting themselves into the air for short, low flights. All three were Silver-laced Wyandottes, a beautiful breed whose white plumage is elegantly lined with black. In flight, however, not even pretty feathers could give their stout bodies a semblance of grace. They hurtled awkwardly forward like overfed wind-up toys, necks jutting out and wings beating frantically to gain a few brief inches of altitude. Wile E. Coyote had a similar look in the old *Roadrunner* cartoons right after he ran off a cliff—his pumping legs kept him airborne for a few seconds, but gravity always won in the end. Watching this spectacle from the porch of the Raccoon Shack, I realized that our chickens might be acting out a scene from their own evolution. Or not, depending on which story you believe.

A running, flapping chicken gets right at the heart of one of ornithology's most divisive questions: ground-up or tree-down? I've posed it to any number of scientists, and they always have a ready answer. The question draws a clear line in the sand between competing camps in the great debate over the evolution of flight. "Ground-uppers" believe that flight developed in fleet-footed theropod dinosaurs, who began flapping and taking short leaps much like our young Wyandottes. The "tree-down" camp is adamant that flight evolved in arboreal creatures as a means to extend their hops from branch to branch. The controversy is irrevocably intertwined with questions about the origins of birds and feathers,

A Silver-laced Wyandotte hen.

and both sides claim supportive evidence from the fossil record—sometimes from the same specimens. The exact sequence of events may be unknowable, but exploring this dichotomy (and a promising alternative) reveals a range of evolutionary possibilities for the origin of that first feathered flight.

The ground-up theory traces its origin directly to Thomas Huxley and his famous papers and lectures linking *Archaeopteryx* with *Compsognathus*. From his suggestion that birds evolved from terrestrial bipedal dinosaurs, it followed that the first fliers were also two-legged runners. Flight makes great sense in this context as a predation strategy, where prolonged leaps, glides, and eventually wing strokes could give theropods and birds sudden access to a sky filled with tasty insects. (Insect flight evolved at least 150 million years before the earliest known birds.) The ground-uppers also note that only a two-legged ancestor would have had its arms free to evolve into independent wings. Most other vertebrate fliers and gliders like bats or flying squirrels descended from quadrupeds who relied on all four legs for locomotion—they couldn't possibly have learned a bird's two-armed flapping without falling face first into the primordial muck. Finally, learning to fly from the ground-up offers a safe environment for the adaptation and refinement of balance, steering, and sophisticated aerodynamics. As many pioneering human aviators (and their survivors) learned, falling or crashing from a low glide carries a lot less risk than leaping from trees or cliffs.

Like the theropod-bird theory in general, the ground-up hypothesis fell out of favor and languished for much of the twentieth century. It reemerged with John Ostrom's reexamination of *Archaeopteryx* and warm-blooded dinosaurs in the 1970s and gained ground as new discoveries underscored numerous similarities between theropods and birds, from wishbones and gizzards to nesting habits and sleep postures. Rick Prum's ideas added fuel to the fire by suggesting that flight adaptations came relatively late in the evolution of feathers. Though he was careful not to preclude a tree-down origin for flight itself, he noted in a 2002 essay that "the major components of the flight stroke had already evolved in bipedal theropods in a wholly terrestrial context." When *Sinosauropteryx* and other clearly feathered bipedal theropod fossils began emerging from the Yixian shales, ground-uppers greeted them as a confirmation and suddenly found their theory pushed into the mainstream.

In spite of this renaissance, however, the ground-up theory still suffers from a major weakness, one key point where its logic falters. It's a lesson that young Fatty, Trouser, and White One learned very well during their abortive sprinting around our orchard: taking off from the ground is difficult. Most birds do it by way of specialized flight muscles and a shoulder joint that give their wings a flexible, powerful downstroke. A series of vigorous flaps provides them both the lift and the thrust necessary for powered flight. Those adaptations appear to be lacking or primitive in *Archaeopteryx*, however, as well as the feathered theropods. (Chickens have the right structure to get airborne, but the wrong body-mass-to-wing ratio for a sustained flight!) There is little doubt that theropods could run and leap, but could that strategy have ever provided the thrust and lift required for takeoff? This is the question that keeps the tree-down theory very much alive.

"There's a certain euphoria in the dinosaur community about having solved the problem," Alan Feduccia told me, but he and other BAND members reject the idea that flight could have evolved in a terrestrial setting. "It's almost un-Darwinian!" he exclaimed. "Look, every group of vertebrates has evolved some kind of flight. There's Draco (a gliding lizard), sugar gliders (a marsupial), frogs, flying squirrels, pterosaurs, bats—all have one thing and one thing only in common. They all evolved flight from the tree down, taking advantage of the cheap energy of gravity."

Dropping from a branch or other high place certainly has been the most prevalent route to the skies, and the list of gliding tree dwellers is impressive. Even ants follow the tree-down model. In the Peruvian Amazon, certain species living in the rain forest canopy have developed flattened, aerodynamic heads and legs that enable them to glide back to their tree trunks after a fall. The tree-down theory puts ancestral birds in that same environment, dashing along limbs and vines and leaping from branch to trunk to branch. An early illustration of the hypothetical species *Proavis*, "before-bird," helped cement this image in our consciousness, and almost every image of *Archaeopteryx* that followed showed it gliding or climbing through such a habitat. It's a setting where any aerodynamic adaptation has immediate and obvious benefits, an important consideration for the incremental process of evolving complex flight. Tree-downers argue that it's hard to envision those intermediate stages developing in a terrestrial ancestor—what use is half a wing for an animal that cannot get off the ground? This is the key element to the tree-down argument: the evolution of flight requires the "cheap energy" of gravity to overcome the lift-and-thrust conundrum so challenging to earthbound chickens and theropods.

Gerhard Heilmann's *Proavis*, a hypothetical bird ancestor that climbed and glided between trees.

The logic is appealing. Even bats, the most nimble nonavian fliers, are helpless on the ground. They require a drop of at least a few vertical feet to reach flight speed, but then their awkwardness gives way to immediate agility. I've watched them emerge at dusk from a roost box on our porch, spreading their wings as they fall and turning immediately at right angles to slip between the deck rails. The air force gives out medals for that kind of maneuver. But while the tree-down theory's cheap gravity is indeed a powerful evolutionary force, feathers complicate things. Feduccia failed to mention something else that all other vertebrate fliers and gliders have in common. Whether marsupial, mammal, amphibian, or lizard, they stay aloft on membranes, thin flaps of skin stretched taut between body parts.

If this book were called *Membranes* instead of *Feathers*, we could spend chapters exploring the more than two dozen independent and unrelated times that membranes evolved for vertebrate gliding and flight. In bats, they connect the fingers of the hand and stretch back to the ankle of the hind foot. For Draco lizards, the membrane extends between flexible, elongated ribs. Pterosaurs had oddly long fourth fingers connected by a fibrous membrane to the hind leg. Wallace's flying frogs soar on giant platterlike feet, their outstretched toes neatly webbed with translucent tissue. Among all flying or gliding animals, only the avian lineage developed the use of feathers for an airfoil. This fact makes Feduccia's argument work even better turned on its head. If time and again the evolutionary answer to tree-down flight is a membrane, then bird feathers are an exception that suggests a different history. With their unique follicles and helical growth, their complex structure and diversity of forms, feathers seem grossly overqualified for the job. Why go to all that trouble when a simple flap of skin would do?

Wallace's flying frogs glide on oversized, broadly webbed feet.

The ground-up–tree-down debate puts many of the familiar figures at odds, but the lines are not entirely clear-cut. As Rick Prum pointed out, a terrestrial origin for feathers does not necessarily mean a terrestrial origin for flight. Surely, it's possible that some already feathered theropod figured out how to climb a tree. Scientists increasingly view the ground-up or tree-down argument as a false dichotomy, proposing new hypotheses that look for a middle path. The discoveries of Xing Xu and his colleagues sum up the situation nicely. On the one hand, fossils like *Caudipteryx* and *Beipiaosaurus* demonstrate that terrestrial bipedal dinosaurs had complex feathers. But *Microraptor* and *Anchiornis* boasted flight feathers on the legs as well as wings, an ungainly arrangement for a running, leaping animal. Xu now believes that such "four-winged" theropods lived in trees and helped bridge the gap from ground to air.

"If you look at all taxa on the line to birds," he explained over the phone, "many flight-related characters evolved in a terrestrial context. The problem is the final step for flight. . . . It needs help from gravity. Arboreal dinosaurs can fill this gap." Though Xu admitted he didn't yet know how the leg feathers on *Microraptor* or *Anchiornis* contributed to flight, he saw them as a critical transitional step. "Four-winged dinosaurs definitely suggest that ground-up cannot explain it alone."

Like so many questions about events in deep time, unraveling the origin of flight is hampered by a woefully incomplete fossil record. Even the wonderful specimens coming out of Liaoning Province provide only glimpses—a few species of theropods and early birds separated from one another by millions of blank years. It's like trying to get the gist of a complex novel by reading one random page from every chapter. Sometimes, the best insights come from studying modern living analogues. Just as Rick Prum looked to the way feathers grow for clues about their evolution, so might the growth of birds tell us something about the beginnings of flight. Could a young bird learning to use its wings unconsciously mimic its own evolutionary history? An ornithologist named Ken Dial asked that question about the Chukar Partridge, an Asian game bird not so different from the young hens running around our orchard. His surprising results formed the basis of a new theory that may be the best bridge yet between ground-up and tree-down.

"We discovered a new behavior, something no one had ever noticed before," he told me by phone from his office at the University of Montana. Ken's voice is low and clear, and he talks with the confidence of someone used to explaining things. In fact, he is—he hosted a series of nature documentaries, called *All Bird TV*, on Animal Planet during the late 1990s. Filming the show took him to field sites around the world, where he saw a pattern in how young pheasants, quail, tinamous, and other ground birds ran along behind their parents. "They jumped up like popcorn," he said, describing how they would flap their half-formed wings and take short hops into the air. So when a group of graduate students challenged him to come up with new data on the age-old ground-up–tree-down debate, he designed a project to see what clues might lie in how baby game birds learned to fly.

Ken settled on the Chukar Partridge as a model species, but he might not have made his discovery without a key piece of advice from the local rancher who was supplying him with birds. When the cowboy stopped by to see how things were going, Ken showed him his nice, tidy laboratory setup and explained how the birds' first hops and flights would be measured. The rancher was incredulous. "He took one look and said, in pretty colorful language, 'What are those birds doing on the ground? They hate to be on the ground! Give them something to climb on!'" At first it seemed unnatural—ground birds don't like the ground? But as he thought about it, Ken realized that all the species he'd watched in the wild preferred to rest on ledges, low branches, or other elevated perches where they were safe from predators. They really only used the ground for feeding and traveling. So he brought in some hay bales for the Chukars to perch on and then left his son in charge of feeding and data collection while he went away on a short work trip.

Barely a teenager at the time, young Terry Dial was visibly upset when his father got back. "I asked him how it went," Ken recalled, "and he said, 'Terrible! The birds are cheating!'" Instead of flying up to their perches, the baby Chukars were using their legs. Time and again Terry had watched them run right up the side of a hay bale, flapping all the while. Ken dashed out to see for himself, and that was the "aha" moment. "The birds were using their wings and legs cooperatively," he told me, and that single observation opened up a world of possibilities.

Working together with Terry (who has since gone on to doctoral studies in animal locomotion), Ken came up with a series of ingenious experiments, filming the birds as they raced up textured ramps tilted at increasing angles. As the incline increased, the partridges began to flap, but they angled their wings differently from birds in flight. They aimed their flapping down and backward, using the force not for lift but to keep their feet firmly pressed against the ramp. "It's like the spoiler on the back of a race car," he explained, which is a very apt analogy. In Formula One racing, spoilers are the big aerodynamic fins that push the cars downward as they speed along, increasing traction and handling. The birds were doing the very same thing with their wings to help them scramble up otherwise impossible slopes.

Ken called the technique WAIR, for wing-assisted incline running, and went on to document it in a wide range of species. It not only allowed young birds to climb vertical surfaces within the first few weeks of life but also gave adults an energy-efficient alternative to flying. In the Chukar experiments, adults regularly used WAIR to ascend ramps steeper than 90 degrees, essentially running up the wall and onto the ceiling. In the wild, precocial species like the Australian Brushturkey use WAIR from the day they hatch and appear to prefer it to flight, even when full grown. As Ken put it, "As soon as you give an animal a three-dimensional environment—a boulder, a cliff, a tree—it will find the best way to climb it."

In an evolutionary context, WAIR takes on surprising explanatory powers. With one fell swoop, the Dials came up with a viable origin for the flapping flight stroke (something gliding animals don't do and a shortcoming of the tree-down theory) and an aerodynamic function for half-formed wings (one of the main drawbacks to the ground-up hypothesis). In terms of feathers, WAIR provides an adaptive role for the transition from a body covering

A young Chukar Partridge uses its wings to ascend steep slopes—did avian ancestors do the same on the path to flight?

to the modern flight feather. Trimming or removing wing feathers from the Chukars consistently caused them to slip backward on ramps—even the youngest partridges were gaining significant thrust from their half-grown plumes. "The feather stuff fits right in," Ken assured me. "What we're finding is in complete and total agreement with Prum's work."

The great appeal of the WAIR hypothesis lies in its ability to provide incremental, adaptive stages between a bipedal, terrestrial ancestor and a flapping, feathered flier. Both the ground-uppers and the tree-downers have reasons to embrace it. For ground-uppers, it shows that even a half wing had potential advantage for a terrestrial theropod and that increased aerodynamics for feathers and wings could lead to greater survival of young birds. For the tree-downers, here was a way for protobirds to start climbing trees in the first place, from which they could have transitioned to true flight. "That's the other half of the story," Ken explained. "WAIR gets the birds up onto an elevated refuge, but they still need to get back down again. So what do they do? Jump and flap!"

Many ornithologists and paleontologists have embraced WAIR as the best flight evolution story to date, but it's not the only new hypothesis out there. I stumbled across another idea at the Wyoming Dinosaur Museum, where one display featured a *Velociraptor*, the voracious predator best known for terrorizing (and steadily reducing) the cast of *Jurassic Park*. The specimen appeared frozen in midchase, leaning into a turn at an alarming angle with its forearms stretched out and clawed feet digging into the sand. In the accompanying description, an illustration showed the beast richly plumed with purple feathers, its wing tips brushing the ground. Though some fossils of *Velociraptor* boast quill knobs—the bony attachment points for flight feathers—no actual *Velociraptor* feathers have been discovered. I called Scott Hartman, the paleontologist and artist who had articulated the skeleton and asked him why he'd given the beast such birdlike wings.

"Steering," he said simply, and explained how theropod dinosaurs lacked the hip flexibility to turn sharply when running at high speeds—an essential skill for a predator. Hartman believes that *Velociraptor* and the other bipedal hunters could have developed aerodynamic wings and feathers to help them turn before they ever dreamed of flapping. "Trees may have played a role in flight," Hartman admitted, "but I can tell you that the animals who climbed them already had wings!"

Decoding every step in the evolution of flight is probably impossible. The story played out over 150 million years in the past, and there's enough ambiguity in the fossils to keep the debate going for decades. As Alan Feduccia told me, "This is the same old stuff I was arguing with John Ostrom about thirty years ago!" But regardless of how birds achieved it, scientists do agree that flight gave the avian line access to a huge range of new evolutionary possibilities. Once airborne, birds spread out and diversified into every available niche, and their bodies continued adapting to life in the skies. Spongy, hollow bones filled with air sacs; toothless, lightweight bills; small, efficient lungs; unidirectional breathing—some of these features have been found in theropods, but all became highly refined in the aerial birds. Flight feathers, too, continued to adapt, the rachis thickening to withstand the strain of flapping, asymmetrical vanes overlapping to work together as a fine-tuned, infinitely adjustable wing. Watching chickens run and exploring the various theories helped me visualize the different ways flight may have evolved. But seeing birds on the wing, with all their grace and casual agility, made me realize that debating ground-up or tree-down hadn't explained something vital. I still didn't understand how bird flight actually works.

CHAPTER EIGHT

A Feathered Hammer

My falcon now is sharp and passing empty.
And till she stoop she must not be full-gorg'd,
For then she never looks upon her lure.

—William Shakespeare,
The Taming of the Shrew (ca. 1590)

Astronauts on the moon don't have a lot of spare time. The Apollo 15 mission marked the first NASA voyage equipped with a lunar rover, and its two-man crew had only eighteen free surface hours to explore the Hadley Rille and the Elbow Crater, take samples from the Genesis Rock, and perform a barrage of groundbreaking tests and procedures. So on August 3, 1971, when Commander David R. Scott stepped in front of a live camera to conduct an unauthorized experiment, he hoped to hell it worked.

"I was going to try it out first," he later recalled, but the mission's tight schedule never gave him the chance. "We ended up just winging it!"

The film, a grainy image beamed straight to ground control and from there to the world, showed Scott in his bulky white space suit holding out two objects, a hammer and a feather. Behind him, the landing module hulked like some dark insect, and beyond that the moonscape stretched off to the black horizon. His voice sounded tinny but clear as he explained what he was doing. "I guess one of the reasons we got here today was because of a gentleman named Galileo," he began, "who made a rather significant discovery about falling objects in gravity fields."

A few seconds later he dropped the hammer and the feather together from shoulder height. They both fell straight down as if pulled by unseen strings, landing at his feet in the gray lunar dust at precisely the same instant.

"How about that!" Scott exclaimed, and the experiment was over. He and his partner returned immediately to their work, packing instruments for the long homeward journey. They couldn't even spare a second to pick up the now famous feather—it remains on the moon to this day.

That short sequence became one of the iconic images from the Apollo space program, and it's still shown regularly in classrooms around the world. Physics teachers use it to illustrate one of the fundamental principles of gravity: falling objects accelerate at the same rate regardless of their mass. Galileo first made this claim in the early seventeenth century, overturning Aristotle's long-held theory that heavy objects fall faster. But demonstrating the law of uniform acceleration alone wouldn't have required a trip to the moon. Legend has it that Galileo tested his theory by dropping various-size balls from the Leaning Tower of Pisa and timing their descent. Scott could have replicated that experiment back home on the launchpad with a brick and a pebble. He needed the moon shot to field-check another of Galileo's hypotheses: in a vacuum the *shape* of an object would be also be

The flight feather of a Peregrine Falcon, dropped on the moon.

immaterial, since there would be no atmosphere to resist its fall. Watching a feather drop in the vacuum of space reminds us that the first ingredient in aerodynamics is "air."

I don't like to take science sitting down, and Scott's work left one obvious trial untested. When Eliza saw me headed for the Raccoon Shack with a ladder, a hammer, and the crow feather I'd found on my walk with Noah that morning, she said, "I think I know what's going to happen." I did too, but I still wanted to see it with my own eyes.

At its peak, the roof of the Raccoon Shack stands eleven feet six inches above the ground, just even with the tallest plum tree in our orchard. I stood there on a sunny afternoon with the feather in my left hand and a hammer in my right, just like Commander Scott. When I set them free, the hammer plummeted straight downward and struck the ground below in just under a second. The crow feather, on the other hand, floated immediately to the side, spun twice in the air, tumbled, caught a breeze, and drifted finally to earth a full six seconds later, twelve feet away from the hammer. It landed quill down, lustrous and black against the grass like a speck of lunar sky.

This exercise accomplished two things. It established the comforting fact that there was a good supply of air around the Raccoon Shack, and it proved beyond doubt that when an atmosphere is present, feathers behave differently from hammers. This latter point is also comforting, particularly to pilots, flight attendants, model-plane enthusiasts, kite surfers, frequent fliers, or anyone else who has depended on airfoils to keep their craft aloft. The feather I used had the distinctive curve and offset rachis of a flight feather, a perfect little airfoil. Its shape and size told me it had come from the crow's left forewing, but it responded to air like a wing in its own right.

Any feather can drift in a breeze; like autumn leaves, they stay aloft by sheer lightness and large surface area. But of the thousands of plumes that cover a bird, only a few dozen wing and tail feathers have the asymmetrical structure of true airfoils. Layered like rows of winglets nested within each wing and tail, these feathers act both individually and in concert to give birds unparalleled control over the nuances of flight.

In his moon experiment, David Scott opted for a falcon flight feather. He chose it to honor his alma mater, the U.S. Air Force Academy (whose mascot provided the plume) and also because *Falcon* was the name of the landing module that brought him to the lunar surface. A third reason made his choice even more perfect.

Perhaps better than any other species in history, falcons embody the aerodynamics of a feathered hammer. Diving from great height to snatch their avian prey from midair, they've been clocked at speeds faster than a locomotive or a Formula One racer. At that rate, the force of their impact is indeed a hammer blow that can shatter the bones of their quarry, driving their talons home deep. Falcons thrive at the extremes of motion. They make their living by outflying other birds, and they've had to develop ways to be faster, more agile, more maneuverable. Watching falcons stoop is a lesson in the possibilities of feathered flight, but they move so quickly, it's hard to make out the details. I needed an insider's perspective on falcon aerodynamics, and I was in luck. The year-round community on our island is small, but it happens to include the world's most famous skydiving falcon, as well as the avid pilot who trained her.

When Ken Franklin jumps from an airplane, he throws Frightful out first. And once both he and the falcon are airborne, he drops a lure, a weighted, aerodynamic bait, that gives her a target to aim for as she dives. Several years ago, Ken added one more thing to the list of items falling from his plane: a National Geographic film crew. Using a tiny modified flight computer attached to her tail, they clocked Frightful diving in a streamlined free fall at 242 miles per hour, a record for animal flight. And with close-up video footage, they were able to see how she did it, stretching her body into a streamlined shape and accelerating downward to do what Ken calls "slipping through the molecules."

I stopped by the Franklin household on a warm midsummer afternoon, and I could hear Frightful's piercing "kek-kek-kek" as soon as I stepped out of the car. "We just let her go today," he said. "She'll fly free for the next three months or so." Hand-reared in captivity, the bird imprinted on Ken and his family at a young age and spends much of her time in their aviary, or even inside the house. As we talked, Ken walked periodically out onto the porch, scanning the sky. It seemed an almost unconscious habit. I joined him just as she appeared, diving low over our heads, seemingly enjoying her freedom but not straying too far from home. I watched her turn, tail fanned and narrow wings flapping as she angled up and landed in a nearby fir snag, calling all the while. "It's risky letting her free. They have accidents all the time," he said with a note of parental worry. "But she'll be fine—she's quite a flier."

Tan, with an athlete's build and hair just beginning to gray, Ken has the faraway gaze of someone used to seeing great distances. He makes his living flying jumbo jets for Federal Express, but it's obvious that his passion lies with the birds. He and his wife, Suzanne, are both master falconers and have raised and trained dozens of birds over the years. Frightful was an apt pupil from the start—alert, loyal, and quick to chase a lure. At thirteen, she has left her skydiving days behind and appears to be enjoying a comfortable retirement in and around the Franklin home. "She's a good hunter, but she's perfectly happy being fed all the time," Ken explained. "A falcon is like anyone else; they'll do the minimum necessary to get by." Then he added, "But they're capable of a lot more."

As we talked, the understatement in that comment became abundantly clear. Ken showed me a photo enlargement of a bird streaking downward in free fall, wings tucked close and body elongated, like a dark teardrop stretched to a fine point. "This is what I wanted to document," he explained. "This body shape. This is what they do when they go into warp drive."

In more than two hundred skydives with falcons, Ken has seen them do amazing things: drafting in the slipstream behind the airplane, revolving to keep their eyes fixed on a spinning lure, or easily keeping pace beside him in free fall. Warp drive happened only occasionally, when they would plummet past him

Peregrine Falcons, by Louis Agassiz Fuertes.

and out of sight, even when he knew he was falling at more than 100 miles per hour. It's what inspired Frightful's speed trials, the National Geographic film, and the IMAX movie that followed.

I asked Ken what role feathers played in a falcon's great speed, and he immediately pulled out a picture of a bird in flight. "Look at the feather edges—they're jagged," he said, tracing a finger along the wing coverts and the contour feathers covering the back. The tips did look uneven where they overlapped, as if some of the barbs were especially long and stiff. "Hawks, eagles, they don't have that," he went on. "It has something to do with airflow, reducing turbulence and drag during their dives. I don't know exactly how it works, but I'm sure that's what is going on." This refrain would become increasingly familiar the more I looked into aerodynamics—feathers enhance bird flight in all kinds of important ways, but the complexity of a living bird wing defies easy quantification.

On his computer, Ken pulled up dramatic photographs of a falcon pursuing a sandpiper at close quarters. The focus was so sharp it captured individual droplets of mud kicked up in the struggle, and every feature of both birds stood out clearly. Frame after frame showed the falcon's feathers changing as it pulled out of its dive, turned sharply, and maneuvered over its prey. The tail fanned wide. Individual wing coverts were lowered and raised. The flight feathers spread, slotted, and changed angle, and in one shot each outer tip was bent sharply upward. "Their flexibility is amazing," Ken said, staring at the screen. "In all the dives I've watched, I've never seen a feather break in flight. Wrestling on the ground with prey, sure, but never on the wing."

That's an impressive statement considering the tremendous structural strain of braking and turning at such high speeds. The flight computer attached to Frightful's tail coverts has revealed some amazing statistics. She once dove after a lure dropped from three thousand feet, accelerating to 157 miles per hour before neatly catching it and pulling up from her stoop a mere fifty-seven feet above the ground. The gravitational force on her body in that moment has been calculated as high as twenty-seven Gs. Fighter pilots risk losing consciousness at anything over nine.

Ken sent me home with a handful of falcon feathers and a lot to think about. Our conversation gave me a new appreciation for the sheer physicality of bird flight, and also its fundamentally three-dimensional nature. In addition to their notorious dives, falcons dart upward and sideways with ease, dance in and out of thermal currents, and hunt their prey in seemingly any direction. "I don't think that vertical and horizontal mean anything to them," he said.

I didn't ask Ken about his motivation or the obvious personal risks he took to jump and fly with Frightful—his drive and curiosity seemed answer enough. But I know he also hopes his work may eventually lead to improvements in aircraft design and that he has collaborated closely with a top engineer at Boeing. Though skydiving with birds may be unique, studying their aerodynamics fits Ken into a long line of aviators who've mixed a healthy dose of ornithology into their innovations. It's a history that puts birds, feathers, and aviation firmly at the heart of one of the hottest trends in the physical sciences.

CHAPTER NINE

The Perfect Airfoil

All she needs is a coat of feathers, and she will stay in the air indefinitely!

—Dan Tate, assistant to the Wright brothers at Kitty Hawk, 1902

My brother is a gifted mechanical engineer. He revealed his career inclinations while still a toddler, by using his toy screwdriver to dismantle a wide range of household furnishings. Removing all the screws from my parents' bed frame, for example, had predictable and rewarding comic results. When I told him that I was writing a book about feathers, he gave me an immediate and emphatic one-word response: "Biomimetics!"

For engineers, physicists, and even chemists, this word is a guiding experimental principle. It means just what it sounds like: the *mimicking* of *biological* structures, behaviors, and processes to create cutting-edge technologies. Biomimicry (or "bio-inspiration") may trace its roots to the first person who followed a game trail to water or used leafy camouflage to stalk an antelope. Certainly, feathers contributed to the field from an early stage. Their addition to arrow shafts marked one of the greatest advancements in hunting and warcraft from the Stone Age to the widespread adoption of firearms. Early hunters could see how feather vanes controlled the flight of birds, and from there it was a short leap to fletching. (By extension, the fins on the massive rocket that launched Apollo 15 were direct technological descendants of the first feathers to stabilize the tail end of a dart.) Though feathers and projectiles have gone hand in hand for millennia, historians of science generally link biomimicry to a deeper human longing, what Wilbur Wright described as "the desire to fly after the fashion of the birds . . . through the infinite highway of the air."

In Greek mythology, the brilliant craftsman Daedalus and his son, Icarus, escaped the clutches of a hostile king on wonderful wings built from feathers, wax, and twine. As they soared away over the sea, Icarus grew bold and overconfident in his flight. Ignoring his father's warnings, he flew higher and higher toward the sun, whose heat softened and finally melted the wax holding his feathers in place. The wings broke apart, and he tumbled to his death in the water below, singed plumes raining down around him. For more than two thousand years, moralists have held up Icarus as a stern lesson on pride and the rashness of youth. But in terms of biomimicry, it was Daedalus who made the greater mistake.

In crafting his famous wings, Daedalus copied only the *appearance* of birds, ignoring the biological and physical processes that made their flight possible. His creation was more chimera than science, capable of flying only in the imaginations of ancient storytellers and their listeners. But that vision was powerful, and for centuries the allure of an Icarus flight both inspired and confounded human efforts to reach the skies. Time and again, aspiring aviators repeated Daedalus's basic error: affixing fanciful wings to their arms without any biological understanding of why this is a good design

The Lament for Icarus, by Herbert James Draper.

for birds but a disastrous one for people. What ensued was a long series of ill-fated leaps off bridges, balconies, terraces, and palace walls, each one involving some hapless inventor strapped into a flapping contraption. The first chapter of Octave Chanute's classic 1894 treatise, *Progress in Flying Machines*, reads like the logbook from an emergency ward, with each would-be aeronaut's attempt followed by a description of his injuries. Mr. Allard (1660)—"grievously hurt." The Marquis de Bacqueville (1742)—"broke his leg." Mr. Letur (1854) "received injuries that resulted in his death." Mr. De Groof (1874)—"killed on the spot." For a Mr. Degen in 1812, surviving his failed flight trials in Paris proved the least of his worries: "On the third attempt he was attacked by the disappointed spectators, beaten unmercifully, and laughed at afterward as an imposter."

After detailing dozens of examples, Chanute concluded that "not only has every attempt of man to raise himself on the air by his own muscular efforts proved a complete failure, . . . there seems to be no hope that any amount of ingenuity or skill can enable him to accomplish this feat." A few years later, French and German engineers bore him out, using equations from the nascent field of aerodynamics to finally prove humanity incapable of self-propelled flight. These observations came at a crossroads for aviation, less than a decade before the Wright brothers finally succeeded in reaching the skies—not by flapping but with the help of fixed aerodynamic wings and a twelve-horsepower motor. The story of that famous first flight, as well as the years of controversy, competition, and heady progress that followed, is a grand tale that's been well told elsewhere. For this discussion, we'll concentrate on the key points where feathers and bird wings touch on the history (and future) of aviation.

Though the Wrights later attributed their triumph more to a good machine shop than any avian influence, both brothers were keen observers of birds. They spent hours watching gulls, vultures, and eagles soar, noting every wing twitch and feather adjustment. Seeing how birds twist the tips of their wings during turns gave Wilbur the idea of "warping," flexing or adjusting a wing's angle to initiate lateral rolls. That breakthrough formed a cornerstone of their patented "three-axis control" system, whose basic principles still govern aircraft steering today. Wilbur's observation and subsequent invention embodied biomimicry, borrowing from nature to advance human technology. Wing warping added to a long history of similar leaps in aviation that date back to one of the greatest innovators of all time.

Leonardo da Vinci dreamed of flying. He wrote that developing a machine capable of reaching the skies would be the one invention to gain him "glory eternal." During the first years of the sixteenth century, Leonardo made careful observations of the Black Kites, larks, and other species common to the Italian countryside, jotting down his ideas in a private notebook that came to be called *Codex on the Flight of Birds*. Leonardo's designs for flying machines included a now-famous illustration of a primitive helicopter, as well as a flapping Icarus-style flier, but the most important pictures in the book may be a series of small birds sketched casually in the margins. They look like pigeons in flight, showing a variety of postures and depicting the birds with lines of air passing under and over each wing. Combined with his text on the "thickness" and "thinness" of air, these pictures make plain that Leonardo had begun to intuit the importance and function of airfoils.

When air meets the front of a bird wing, it has a choice: take the low road, or take the high road. Both paths reach the other side, but they travel by different routes, at different rates, and through surprisingly different conditions. The angle and speed of the wing determine how much air is deflected downward, creating lift by increasing air pressure below the wing while reducing it above. It's Newton's third law at work: every action has an equal and opposite reaction. This effect is familiar to anyone who has held their arm out a car window and felt the wind force their hand up as they tilt it to cup the air. Shape matters, too. Seen in cross-section, a bird's wing has a curved top, a thick leading edge, and a long, tapering tail, like a comma tipped on its side and stretched thin. Airflow hugs that curved upper surface and exits behind the wing as downwash, further reducing the air pressure above and adding additional lift. This is also easy to test. Dangle a piece of paper in front of your lips and blow across the curved top surface—you will see the paper rise from below, pushed upward as your wind reduces the air pressure above the page.

Leonardo was particularly well suited to understanding aerodynamics, having already studied the way water moved in streams, around obstacles, and through tubes of various widths. He was the first person to grasp that air and water moved by the same principles and is considered the father of the combined field of fluid dynamics. Though no one thinks of Leonardo as the father of ornithology, Rick Prum once made a study of his ornithological observations and noted how close he had been to the modern understanding of bird flight. If he knew what we know now, Prum observed, "he would have cursed himself for not figuring it out!"

Had da Vinci published his findings, he might have spurred a great Renaissance interest in aerodynamics and advanced the goal of human flight by centuries. But the *Codex* languished unread in private hands, forcing other thinkers to rediscover Leonardo's ideas on their own. Detailed notions about birds and airfoils didn't emerge again until the late 1800s, with a pair of German brothers who "devoted the greater part of our immature nature studies to watching our friend the stork."

Otto and Gustav Lilienthal quickly progressed from observation to action, building their first flying machines while still in their teens. Though early attempts involved sewing together "all the feathers which were obtainable in our town," they soon learned that the overall shape of the wing, the airfoil, was more important than replicating a bird exactly (the error that Daedalus made). They experimented with a wide range of airfoil shapes and recognized the importance of a curved upper surface. Their designs took on a distinctive look—heavily battened spreads of canvas surrounding a lone upright pilot, who controlled the craft by shifting his body weight from side to side. There was no motor. The Lilienthals made use of headwinds and updrafts, learning to

Otto Lilienthal displays one of his many flying machines, 1894.

soar with wings unmoving, just like the great white storks they sought to emulate. Known as the Glider King, Otto became the world's first truly famous aviator, completing more than two thousand successful flights—some longer than a thousand feet—before a crash in 1896 ended his life. Orville and Wilbur Wright cited his exploits as their primary inspiration and based much of their early work on airfoils and equations from his book, *Birdflight as the Basis of Aviation.*

In many ways, the Wright brothers began exactly where the Lilienthal brothers had left off. But while they learned much from their predecessors' triumphs, they learned even more from their failures. Otto's fatal accident taught them that control in the air was ultimately more important than simply getting off the ground. The Wrights built a series of gliders that focused on steering and quietly began breaking all of the Lilienthal records for distance, speed, and duration. When it came time to add power to their flier, they had another forerunner to look to, one who also found inspiration in the natural airfoils of the birds.

While the Wrights and Lilienthals achieved unquestioned technical success, it was up to a Frenchman to imbue aviation with a sense of style. Clément Ader made his fortune in the telephone business, introducing and adapting Alexander Graham Bell's invention to an eager Parisian market. He then devoted himself to flying machines, investing the modern equivalent of millions of his own dollars into a series of elegant steam-powered aircraft. Like his contemporaries, Ader believed that the key to flight lay in the effortless soaring abilities of large birds. Instead of storks, however, he focused on vultures and once traveled to the hostile interior of Algeria, disguised as an Arab, to view African species in the wild. His designs looked more like bats in the end, but Ader's greatest contribution lay less in wing shape than in means of propulsion.

Though it rose only eight inches into the air and couldn't be steered, Ader's Éole claimed rights to the first self-propelled manned flight in human history in 1890, thirteen years before the Wright brothers at Kitty Hawk. The plane's small steam engine powered a graceful propeller made up of four giant feathers carved from wood. Ader had recognized that flight feathers, like wings, were perfect airfoils and that the same aerodynamic principles that provided lift could provide thrust if the vanes were mounted vertically and spun at great speed. That separation of thrust and lift became a signature of virtually all future airplanes, and while the Wright brothers transformed propeller design into an exact science, Ader's whimsical feathery blades lived on in a burgeoning side industry. As late as the 1950s, model-airplane enthusiasts still built *The Featherfly* and other designs that called for plumed props.

In the modern era of jet propulsion and in-flight DVD rentals, feather-shaped propellers seem decidedly quaint—something

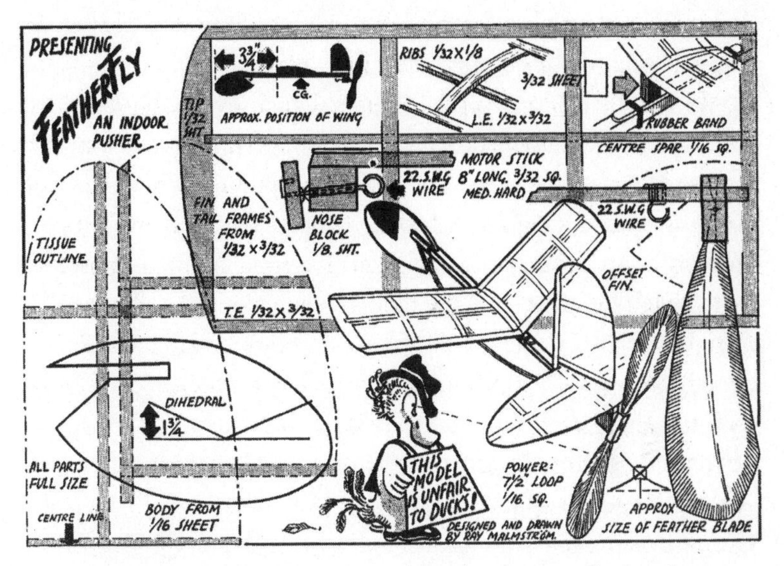

Using feathers for propellers persisted in model airplane designs until at least the 1950s.

you might find in an old textbook or a dusty side room at the Air and Space Museum. After the Wright brothers, conspicuous biomimicry declined in aviation, replaced by equations, wind-tunnel experiments, and, eventually, computer simulations. Few aeronautical engineers today spend much time outdoors stalking storks, hawks, or Algerian vultures. With the principles of flight and steering established, advances during the twentieth century basically focused on improving the components of the Wright Flyer—stronger engines, better airfoils, more responsive steering, larger wings and payloads. This is the evolution of technology: a large leap followed by a thousand tiny refinements.

Within a few decades, aircraft designers had essentially perfected the form. The Cessna 172 is the world's most popular airplane, a reliable and efficient four-seater little changed from its original 1955 design. Boeing's 737 was introduced in the 1960s and remains the world's top-selling jetliner. When innovation turns to impasse, technology often goes back to its roots for inspiration. The need for more efficient, maneuverable, quiet aircraft has brought biomimicry back into vogue, as top engineering labs reexamine the many nuances of feathers and bird flight. But where nineteenth-century aviators had only field glasses and a notebook, modern engineers can watch birds with high-speed cameras, laser range finders, digital modeling software, and other tools that delve beyond mimicry to reveal the fundamental wonders of feathered flight.

A driving force behind the aviation industry's "second look" at bird flight lies in fuel efficiency. Spiking oil prices in 2008 contributed to a slew of airline bankruptcies, including such major carriers as Japan Airlines, Frontier, Skybus, and Aloha. Surviving companies are eager for ways to reduce fuel consumption in a world of ever-shrinking petroleum reserves. Lighter airplanes offer one option, but replacing aluminum with composites involves a complete redesign, with all new materials and manufacturing techniques. The first plane to go this route, Boeing's 787 Dreamliner, shows great promise but recently missed its seventh delivery deadline and remains mired in testing and development, three years overdue. Some engineers believe that quicker fixes lie in how a bird manipulates its wing feathers, a masterful display of what might best be called *airflow management*.

Textbook diagrams of airfoils regularly leave out a critical detail: turbulence. Air passing around a wing never actually moves in the smooth lines of illustration—it swirls and eddies in complex patterns that change constantly with every subtlety of temperature, air pressure, wind speed, wing shape, and angle. There are layers of air dragged along with the wing, vortices tumbling above its surfaces, and spirals jetting off the tips. The process is far too complex for any drawing, but understanding it is critical to understanding drag, the natural resistance to a wing's forward motion. Any reduction in drag increases flight efficiency, offering immediate fuel savings to airlines. And no one reduces drag better than birds.

If you've ever flown in a window seat, you may have admired the silvery shine of an airplane wing and watched its several flaps raise and lower at various times during the flight. It's a precise and beautifully designed instrument, but must look terribly crude to a bird, whose own wings can flap and flex, extend and contract, spread, narrow, tuck, and twist, responding instantly to ever-changing conditions. Taken together, the overlapping flight feathers create a single dynamic airfoil. But they can also move independently and are themselves shaped like airfoils, acting as individual winglets within the greater whole. Vultures, eagles, and other soaring birds use small adjustments of their spread wing-tip "fingers" to manipulate air currents or change speed and orientation, and all birds utilize feather movements to instinctively alter the turbulence patterns around their wings. Slots can be opened or closed to direct air between primaries; covert feathers can be raised or lowered like tiny flags—the possibilities are endless.

Teasing apart these intricacies challenges even the most advanced computer models, but engineers have already learned that adding artificial "winglets" to the tips of airplane wings can mimic the efficiency of a soaring raptor. Passenger jets retrofitted with winglets have seen their fuel use drop by as much as 6 percent, a substantial savings considering that a fully loaded 747 can burn through a gallon or more every second. Now in common use, these small vertical fins have saved the airline industry billions of dollars in fuel costs. A potentially more lucrative lesson can be summarized in one unexpected word: *fuzziness*.

Photographs of birds in flight often show splayed and uneven flight feathers, or coverts lifted at sharp angles above the wing—like Ken Franklin's amazing pictures of a falcon catching shorebirds. Engineers now believe this intentional "roughening" of the wing surface may substantially reduce turbulence and drag. A fully feathered jetliner is probably out of the question, but simulations suggest that just covering the wings with simple bristles could improve flight efficiency by as much as 15 percent. Typically, air passing over the surface of a wing (or any airfoil) breaks apart into tiny eddies that pull away from the surface, a form of turbulence that results in additional drag and pockets of dead air directly behind the wing. When bicyclists tuck into the slipstream behind a lead rider, they're taking advantage of this principle—saving energy by riding in a low-pressure, low-turbulence position. It's counterintuitive, but rough surfaces can help reduce drag by manipulating the formation of eddies and keeping them close to the airfoil surface. Years from now, you may peer out an airplane window and see a fuzzy wing, each bristle the manufacturer's best approximation of a feather.

Managing the airflow around wings can have the added benefit of substantially reducing aircraft noise, an important consideration for anyone living on a busy flight path. When owls pass overhead, their eerily silent wing strokes seem otherworldly, and they've long been linked to mythologies of the spirit realm. But there is nothing supernatural about an owl's flight—their wings simply part the air in a different way. Owl feathers feature barb extensions on the leading and trailing edges that reduce turbulence over *individual feathers* as well as the entire wing, increasing efficiency but, most important, muffling the sound of their approach. This stealth gives them a key advantage over the wary and sharp-eared mammals and birds they prey upon. (In a tidy and satisfying piece of evolution, the Scops Fishing Owl lacks these feather modifications—there's no need to fly silently when your prey is underwater and can't hear you coming!) The promise of quieter airplanes makes owl feathers an intriguing model for commercial carriers, whose takeoffs and landings at urban airports are dictated in part by sound levels.

The complexity of airflow around bird wings (here a Northern Cardinal) continues to inspire aeronautical engineers.

Reducing drag and noise offers immediate and practical applications of biomimicry, but the true dreamers in aviation want to rethink the entire notion of an airplane. After Otto Lilienthal's untimely death, his brother, Gustav, continued to pursue their ultimate vision: a flying machine powered not by propellers or jet turbines but by flapping, birdlike wings. For nearly forty years he designed increasingly ambitious (and sometimes outrageous) aircraft at a small hangar outside Berlin. Known as "ornithopters," from the Greek for *bird wing*, none of these craft ever left the ground, and Gustav became something of a pariah in the burgeoning era of fixed-wing planes. Nevertheless, various engineers have picked up and carried his torch over the years, and their efforts may finally be bearing fruit.

The RoboSwift is an unmanned surveillance plane built in the Netherlands and inspired by the graceful flight of its avian namesake. Though still powered by a propeller, its wings sweep backward to dive and turn, and they extend for efficient soaring. Another Dutch team recently launched the DelFly II. It too is a plane intended for unmanned surveillance, but one whose thrust and lift come entirely from its madly flapping wings. This little craft proved so birdlike it was attacked by a male starling and driven to the ground on one of its first outdoor missions. The prize for a successful manned flight, however, goes to Project Ornithopter at the University of Toronto. Their full-size craft launched briefly skyward under its own power in 2006, to the apparent amazement of even its pilot, who emerged from the cockpit after a hard landing shouting, "It flew! It flew! It flew!" Though the flight lasted only fourteen seconds, that puts it in good company—the Wright brothers' first jaunt lasted only twelve.

Whether by ornithopter, glider, or fuzzy jet, human attempts at flight will always be conflated with our understanding of birds, their feathers, and the incredible aerodynamics of their wings. The yearning for an Icarus-like ride lives on in our technologies, but even more so in our myths. From Superman to Peter Pan to the *Matrix*, we imbue our greatest heroes with the ability to soar like a bird. Skydiving, hang gliding, kite surfing, and other extreme sports may offer a glimpse of this feeling, but our closest approach to bird flight will always lie in dreams. Though Karl Jung and Sigmund Freud disagreed about the symbolism of dream flight (transcendence versus sex), their theories really only proved that neither was a bird-watcher. Wilbur Wright was probably closer to the truth in a speech to the Aéro-Club of France in 1908, when he chalked it up to an age-old envy, the wish to replace our wearisome, earthbound trudging with something free and pure: "I sometimes think that the desire to fly after the fashion of the birds is an ideal handed down to us by our ancestors, who, in their grueling travels across trackless lands in prehistoric times, looked enviously on the birds soaring freely through space at full speed, above all obstacles, on the infinite highway of the air."

Fancy

As long as there are women in the world there will always be feather buyers.

—Parisian feather merchant, *Cape Times* (1911)

CHAPTER TEN

The Birds of Paradise

I looked with intense interest on those rugged mountains, retreating ridge behind ridge into the interior, where the foot of civilized man had never trod. There was the country of the cassowary and the tree kangaroo, and those dark forests produced the most extraordinary and the most beautiful feathered inhabitants of the earth, the varied species of Birds of Paradise.

—Alfred Russel Wallace approaches New Guinea, *The Malay Archipelago* (1869)

Like fishermen, bird-watchers have been known to exaggerate. But unlike fishing, where the tall tales develop afterward and can be blamed on memory or cheap beer, bird-watching hyperbole begins with the very name of the activity. For the most part, the phrase should really be *bird identification*. Though we may set out intending to watch birds, we don't often meet the definition: "to observe attentively; typically over a period of time." Our binoculars seem to have minds of their own, swinging quickly away as soon as we can put a name on the species: House Wren,

Laughing Gull, flicker, crow, chat. It's a dangerous trap, because the true wonder of birding lies in the watching, soaking up the fine details of plumage, behavior, and habit. Even common birds do uncommon things, and every sighting is worth more than a glance and a tick on a checklist. I try to be vigilant, but on the one day I saw a bird of paradise, I totally blew it.

It was mid-March in Australia, and the coast road north from Port Douglas had just reopened after heavy rains. Some fellow biologists and I were on a break from field studies, hitchhiking ever northward toward Cooktown and the fabled Cape York Peninsula. Reaching within one hundred miles of New Guinea, Cape York's eastern shore mirrors the island's tropical climate, a strip of rain-forest green airbrushed on the edge of Australia's dry, dry interior. It's a place where kangaroos climb trees and giant blue-headed cassowaries roam the forest floor like theropods reborn, their toes armed with six-inch claws and their heads topped by sharp, bony crests. Though it may sound like a busman's holiday for a bunch of biologists on vacation, who could resist the allure of such unique creatures?

We made it as far as Cape Tribulation, a beach enclave nestled entirely within Daintree National Park. At the time it boasted eighty residents, a backpacker hostel, a takeout fish joint, and a convenience store located in someone's living room. "The rest," my journal reads with satisfaction, "is forest and beach and reef." It was there, on a muddy road through the rain forest, that I spotted a glossy black bird perched high up in the canopy. Its bill curved gently downward, and its iridescent green throat flashed in the morning sunlight, giving the game away. "Victoria's Riflebird," I noted and moved on, eager to see as much as possible in the short time we had to explore.

In my haste, I passed up the opportunity to witness one of nature's greatest displays, a living lesson in the sexual allure of plumage. Had I truly watched that bird, I might have seen him puff out his brilliant green throat and arch his wings to form a perfect ebony circle surrounding his head. He would have tilted his beak to the sky and opened it wide, exposing the vivid golden skin of his mouth as he sang out in loud, rasping notes. With luck, a tawny brown female would have landed facing him, bobbing her head, raising her own wings, and joining in an elaborate swaying dance back and forth along the branch. Nature films can't help but set this flamboyant spectacle to tango music, and it's a perfect fit. Instead of this, my experience of the Victoria's Riflebird consisted of a single checkmark beside a picture in a field guide.

Like the other strange fauna of Cape York, riflebirds have more in common with tropical New Guinea than the Australian Outback. They belong to the island's most celebrated avian family, Paradisaeidae, the birds of paradise. While my treatment of paradise birds was inexcusably cursory, other naturalists have suffered the exact opposite problem. In 1858, Alfred Russel Wallace was obsessed with them.

On July 1 of that year, Wallace was New Guinea's sole European resident, living in a twelve-by-twenty-foot hut in the village of Dorey on the island's rugged northwest coast. He had just spent a rewarding day collecting beetles (ninety-five species) but hoped the rains would ease so he could resume pursuit of his true goal: "the most beautiful feathered inhabitants of the earth." Though Dorey proved a disappointing location for paradise birds, July 1, 1858, still marked Wallace's single most significant contribution to science. Unbeknownst to him, the esteemed scholars gathered that day for the monthly Linnean Society meeting in London would hear an essay of his, read together with papers by Charles Darwin, that introduced the theory of evolution by natural selection to the world.

The story is legendary. Where cautious Darwin ruminated on his hypothesis for two decades, Wallace's epiphany came in a malarial fever and he wrote it up in two days. The men maintained a regular correspondence on scientific matters, so it was natural for Wallace to send his essay to Darwin for comment. Mailed from an outpost in the Moluccas, it arrived at Darwin's country estate like a thunderbolt and finally jolted the cautious scholar into action. At the urging of colleagues, he carefully established precedence with the Linnean Society papers and then published *On the Origin of Species* the following year, changing the face of science forever. Relations between the great naturalists remained friendly, and Wallace later dedicated his memoirs to Darwin with "deep admiration for his genius and his works." But Darwin's prominence would forever put Wallace in the odd position of being famous for his lack of fame, less known for his ideas than for the want of credit he received for them.

At the time, however, these events passed Wallace by completely. He remained in the Malay Archipelago for three more years, enduring fevers, isolation, and near starvation as his collections grew to a staggering 125,660 specimens, including more than 1,000 species new to science. There were beetles, orangutans, butterflies, kangaroos, snakes, bats, shellfish, and a flying frog, but among all these, nothing captured his imagination more than birds of paradise. Reading his description of their "dancing parties," it's easy to see why:

> On one of these trees a dozen or twenty full-plumaged male birds assemble together, raise up their wings, stretch out their necks, and elevate their exquisite plumes, keeping them in a continual vibration . . . so that the whole tree is filled with waving plumes in every variety of attitude and motion. . . . The wings are raised vertically over the back, the head is bent down and stretched out, and the long plumes are raised up and expanded till they form two magnificent golden fans striped with deep red at the

> base, and fading off into the pale brown tint of the finely divided and softly waving points. The whole bird is then overshadowed by them, the crouching body, yellow head, and emerald green throat forming but the foundation and setting to the golden glory which waves above.

With these words, Wallace began to unravel a mystery that had puzzled naturalists for centuries. His "dance party" was what biologists now call a *lek*, a type of communal display where groups of males gather to pose and posture in an intense competition for mates. The word derives from a Swedish verb for "play," but there's nothing playful about lekking males. The quality of their performance proves more than status on the dance floor; it determines just who among them will reproduce and who will remain an evolutionary wallflower. Species that lek often develop exaggerated characters or behaviors specific to the mating ritual. Certain antelope, a fish, and even a tiny white moth are known to do it, but lekking finds its greatest expression in birds of paradise, and it helps explain why they've developed the most varied and colorful plumage in history.

Wallace's "magnificent golden fans" belong to the male Greater Bird of Paradise. They include hundreds of elongated contour feathers that trail from its flanks in fiery streamers, stretching two times its body length or more. From head to tail the bird's color scheme is so ornate it's hard to follow. A guidebook description includes no fewer than fourteen distinct feather shades, from "warm sepia" and "walnut brown" to "maroon," "orangy yellow," and "vinaceous pink." And the Greater Bird of Paradise is just the beginning.

Taxonomists recognize forty-two species of paradise birds, each with a unique suite of elaborate displays and courtship regalia. Some birds puff out ebony emerald-topped feather skirts and waggle

Plume hunters stalking male Greater Birds of Paradise in the Aru Islands (from Alfred Russel Wallace's *Malay Archipelago*).

the hula. Others dangle upside down from branches or juggle iridescent-green feather coins suspended on long filaments above their heads. There are turquoise and purple ruffs that extend sideways like skinny bow ties or inflate like a lion's mane. The King of Saxony Bird boasts two half-meter feathers jutting from the back of its head, each emblazoned with fifty sky-blue flags that sway and bounce seductively in front of prospective mates. The bent, naked tail quills of the Twelve-wired Bird look so bizarre, the first specimens brought to Europe were long dismissed as fakes. To page through a book on birds of paradise is to be astounded, to feel that nothing is left to exaggerate. From Wallace's time onward, these birds have embodied the extremes of an evolutionary process that Darwin dubbed "sexual selection."

Published in 1871, *The Descent of Man, and Selection in Relation to Sex* marked Darwin's second major contribution to evolutionary theory. Though he intended the book as a treatise on human origins, he ended up devoting more than half of it to the idea that breeding behaviors are a potent evolutionary force. He seemed surprised by the result, as if the importance of the second topic had grown with the writing of it: "Consequently, the second part of the present work, treating of sexual selection, has extended to an

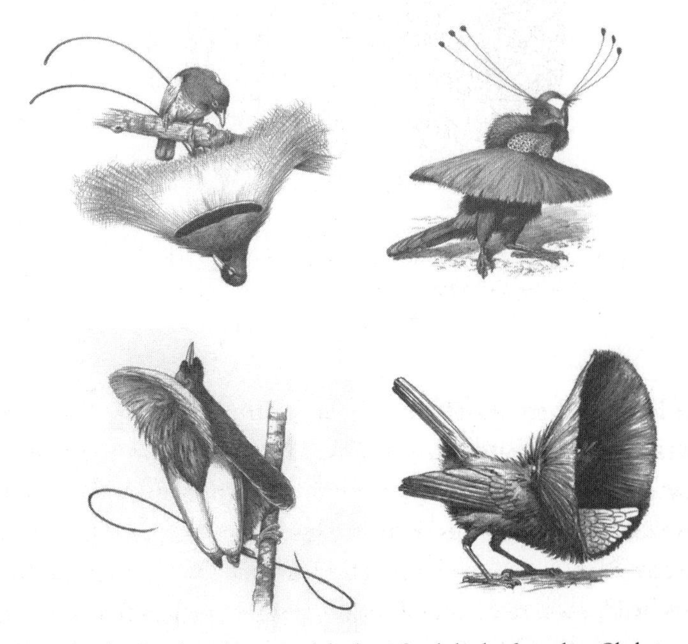

The elaborate breeding plumage and displays of male birds of paradise. *Clockwise from upper left:* Blue Bird of Paradise, Western Parotia, Superb Bird of Paradise, and Magnificent Bird of Paradise.

inordinate length, compared to the first; but this could not be avoided." Indeed, though Darwin's treatment of human evolution showed his typical thoroughness and took its place next to classic works by Huxley and the eminent German biologist Ernst Häckel, it was sexual selection that proved the book's most novel and enduring idea.

Darwin argued that competition for mates constituted a distinct evolutionary process, resulting in dramatic differences in body size and ornamentation between males and females (what biologists call "sexual dimorphism"). Where his law of natural selection governed the struggle for survival, sexual selection applied to the struggle "for the possession of the other sex." In this way it operated outside the strict bounds of adaptation and fitness, helping to explain strange, exaggerated features with no other discernible purpose. Chief among his examples were the extremes of avian plumage. He devoted four chapters to birds and their feathers, from peacocks and pheasants to hummingbirds and hornbills, as well as the various birds of paradise he learned about from his correspondence with Wallace.

Countless experiments and field studies later, evolutionary biologists now distinguish between two basic forms of sexual selection. Direct competition for dominance (typically among males) leads to the evolution of large body size and weaponlike appendages. Picture the massive tusks of a male elephant, rams bashing their horned heads together, or the broad chest and sharp canines of a silverback gorilla. Called intrasexual selection, this is the clumsy, violent modus operandi typical of mammals—males whack one other around in some fashion until a victor emerges in a "winner-takes-all" bid for mating rights. For a more nuanced story, bird lovers and feather enthusiasts can turn with satisfaction to the version that results strictly in ornamentation, the evolution of fancy.

Birds practice what is called intersexual selection, a process that relies on one of nature's fundamental principles: females are choosy. It turns out that this axiom applies to more than ice-cream flavors, furniture, and the color of the baby's room—in fact, it takes on evolutionary significance in decisions surrounding mate choice. Putting females in the driver's seat creates an evolutionary imperative for male birds to stand out from the crowd. Some do it by singing or developing elaborate flight displays and courtship dances, but the most dramatic outcome of sexual selection in birds is colorful plumage. Without it, feathers and arguably birds themselves would never have become so diverse, this book would never have been written, and generations of bird-watchers would never have waxed rhapsodic about the beauty of a male in its breeding finery.

Wherever you see dimorphism in plumage, sexual selection lies at the root of it. But it's important to remember that natural selection also plays a role. Female "drabness" represents its own finely tuned strategy for reproductive success. Finding a good man is rarely a limiting factor—breeding opportunities abound wherever males are displaying, singing, defending territories, or otherwise showing off. For the female to pass on her genes, she must be hidden while sitting on a nest and remain inconspicuous while foraging to feed her hatchlings. In this context, natural selection favors camouflage, the cryptic browns and streaked tans common across a spectrum of females, from ducks to finches to warblers.

Sexual selection abounds with subtleties, but two main schools of thought describe how it actually works to favor gaudiness. In the "good genes" theory, better ornamentation signals increased health and vigor. Males who can afford the high energetic cost of producing and maintaining elaborate feathers (as well as avoiding predators while being so obviously adorned) must be inherently stronger. Females pick them as the best choice to father healthy offspring, and the plumage becomes increasingly embellished over time. Proponents of "runaway selection," on the other hand, suggest that ornamentation is essentially random, not linked to any underlying favorable traits. Also known as the "sexy son" or "fashion icon" theory, this model calls for the gross exaggeration of any idiosyncrasy favored by females—beauty for beauty's sake.

The good-genes concept draws support from a range of memorable studies. Female Barn Swallows, for instance, consistently choose males with the longest tail streamers, the same males that show a significantly lower parasite load than their competitors. In Barn Swallows, long tails accurately advertise good health. For female Bobolinks, the sexiest males carry on the lengthiest display flights, flashing their pied wings and bright-tawny napes. Not coincidentally, these males have the highest fat reserves and highest success rates in fledging their young. In truth, the two models probably overlap, with such "honest advertisements" of health becoming magnified beyond measure in a "sexy son" scenario. But to illustrate classic runaway selection, textbooks invariably point to the birds of paradise, a group so overwrought in plumage that they were once thought otherworldly.

When Wallace finally returned to England, he brought in his luggage two male Lesser Birds of Paradise, kept alive on a diet of bananas, rice, and insects. Whenever the steamer called at a new port, he raced ashore to find food for the birds and later described a stop at Malta with satisfaction: "I got plenty of cockroaches at a bake-house, and . . . took with me several biscuit-tins' full, as provision for the voyage home." The birds survived and fetched a good price from the Zoological Society of London, the fellows of which were so grateful, they also granted Wallace free admission for life. In an age fascinated with natural history, these were the first live birds of paradise ever presented for public display. They drew large crowds and helped to finally dispel lingering myths about the origins of these enigmatic birds.

For more than three centuries, Europeans had marveled at the occasional feathered skins that returned home with explorers and travelers from the Far East. As remarkable as their rich russet feathers and golden plumes was the fact that the birds appeared to lack either wings or feet. Purchased from Malay traders, who themselves had never seen a live one, these specimens were accompanied by an enduring legend. The Malays called them *God's birds*, the *birds of paradise*, divine creatures that inhabited the lower realms of heaven where they drifted in the ether with no need to flap or land. They were not hunted—people found them only after they had died and fallen to earth.

In truth, the New Guinean tribesman who actually hunted birds of paradise simply knew their market, discarding useless wings and feet in favor of the most valuable plumes. Prized as trade items, paradise feathers occupied a luxury niche in the regional economy, appearing in the regalia of chiefs, nobles, and kings as far away as Thailand and Nepal. The paradise story persisted, however, and naturalists of the day took it seriously, proposing ever more outlandish behaviors to explain the ecology of these mythic birds. What did they eat in heaven? Dew and ambrosia. How did they sleep? Tied off to branches by their long tail feathers. Where did they nest? On the backs of the males, where females deposited their eggs and sat upon them in a comfy, feathered hollow.

Wallace countered these fables with careful observation, Darwin supplied the theory, and generations of field biologists (including such luminaries as Ernst Mayr and Jared Diamond) have filled in the details. But still these wildly plumed birds retain an aura of mystery. In spite of their fame, many remain poorly known, hidden away in New Guinea's rugged, mountainous interior. During his eight-year expedition, Wallace saw only five species in the wild, and no single person has ever observed all forty-two. Their feathers remain a valuable trade item in New Guinea, where tribesmen still decorate themselves with plumage for ceremonial village gatherings. Called "sing-sings," these events provide an important venue for young men to display their status to potential wives, in a conscious imitation of birds at a lek. Women and their families can then properly evaluate the contenders based on the number and quality of valuable plumes that each was able to muster. Whether owned, borrowed, or even rented for the occasion, the feathers of paradise birds represent an essential symbol of male stature.

Men of the Obena People of Papua New Guinea performing at a "sing-sing" in December 1991. The men are wearing feathers of at least six species, including the King of Saxony and Superb Birds of Paradise.

People have co-opted bird feathers for adornment throughout human history, a theme we will explore in detail in the following chapter. But birds of paradise represent the extreme in ornate plumage, and there is a human analogy worth mentioning here. Where New Guinea harbors its birds in dense jungles, another paradise exists where human "birds" display in a wholly different, yet eerily similar, fashion.

Paradise, Nevada, lies at the center of a broad, flat-bottomed valley that stretches from the Sierra Nevada west to Spring Peak. The U.S. Census Bureau calls it a "census-designated place," the catchall phrase they use for any populated area that falls just outside a city's actual boundary. Gamblers and tourists from around the world have another name for Paradise. They know it as the Las Vegas Strip.

Flying in, my plane banked low over craggy foothills and dry washes, a countryside of sun and dust, streaked with cloud shadows. It held a certain desolate beauty, the bare rock and sand textured with specks I knew to be creosote bush, mesquite, and sage. I once spent several highly enjoyable weeks studying native bees in a similar landscape nearby, but this trip would be different. My luggage contained no tent or sleeping bag, no insect net, no dissecting microscope, not even a hand lens. Instead, I had washed my one presentable pair of trousers before leaving home and scoured the closet for a shirt with buttons.

The plane descended farther, and Las Vegas suddenly appeared, an unlikely grid of stucco and asphalt sprawled across the desert, decorated with shopping malls, brilliant golf courses, and the tiny blue jewels of countless backyard swimming pools. Above it all, the Strip rose like some sort of hotelier's hallucination. I could see a faux Sphinx and an Eiffel Tower. There were imperial palaces, volcanoes, pirate ships, Venetian canals, and a pyramid. All in all, it looked like an unlikely habitat for a field biologist.

If the birds of paradise embody runaway selection, then the Vegas Strip is runaway leisure, runaway entertainment. It started at the airport, where a huckster boarded our shuttle to point out the sights and hawk discount show tickets, restaurant deals, and time-share condominiums. We passed a hotel called the Mirage, which seemed an apt name for the whole place. The fact that most visitors to Las Vegas never actually set foot in Las Vegas is an appropriate part of the illusion. As the van inched through traffic, I declined two steak-house coupons and a Carrot Top ticket. I had come to town with one goal in mind.

Hours later I found myself in a crowded theater at Bally's Hotel and Casino, sitting between a mother and daughter on holiday from rural England and a large group of Japanese tourists. Like me, they were all first-timers. Everyone's small talk involved aspects of the same theme: the hotels and how darn big they were. Suddenly, the house lights went dark, there was an expectant hush, and then the stage exploded with showgirls.

Extravaganza is defined as an elaborate entertainment, while a *spectacular* is something performed on a large scale, with striking effects. These words don't begin to describe the lavish blast of light and music that poured out over us. More than seventy-five dancers filled the stage, ascending sequined staircases, spinning on platforms, and parading around a mirrored dais. Their costumes glinted with rhinestones, and some of the girls were topless, but nothing in their appearance could take away from the feathers. Countless white and yellow plumes rose up in great fanned tails and headdresses towering five feet above each dancer's head, while boas and trains descended down their backs in golden waves, swaying with every step.

Minutes passed before I realized that my mouth was literally hanging wide open. I closed it with a snap that would have been audible if not for the volume of the opening number (appropriately titled "Hundreds of Girls"). But my amazement stemmed from more than just the show's over-the-top production: I had expected bright lights and loud music, I knew there would be showgirls, and I knew they would be wearing feathers. What left me stunned was how much they looked like birds displaying at a lek. Just as the hotels of Las Vegas replicated famous cities and landmarks, so too were the showgirls engaged in a grand mimicry. Festooned with golden plumage "in every variety of attitude and motion," they moved and swayed exactly like Wallace's birds of paradise, dancing in their rain-forest treetops.

Known as "Jubilee!" (always with the exclamation point), this blowout pageant has shown continuously, except for Super Bowl Sundays, since it first debuted in 1981. It's one of the longest-running revues in Las Vegas history, and for anyone interested in feathers, it provides a crash course in the extremes of *human* plumage, how people have co-opted bird feathers for their own sensual displays. Like the shows at Paris's famed Moulin Rouge, Jubilee! represents the culmination and continuation of a long tradition of feathered fashion and the intricate craftsmanship that goes with it.

"It's their elegance, the softness . . . ," he paused, searching for the right words, "and what they imply." It was the following morning, and I had just asked the head of Jubilee!'s costume shop why they used feathers on virtually every outfit. A twelve-year veteran of the show, Marios Ignadiou spoke in a soft British accent and carried himself with the stylish elegance I learned to expect in the fashion world. He patiently showed me how each piece was constructed, how a rachis could be replaced with stiff wire and bent to any shape, and how plumes could be clipped, brushed, fluffed, and sewn together in myriad combinations. A

The elaborate plumage of a Las Vegas showgirl reflects a long history of feathers in fashion.

single headdress might use two thousand feathers and weigh more than twenty pounds. "And it's all handwork," he told me. "Every costume is done entirely by hand."

Each person I talked to backstage seemed to have a favorite from among the one thousand costumes used in the show, and they all spoke of the featherwork with an affection that bordered on rapture. My notes record how Marios, the dancers, and the wardrobe technicians touched the feathers—stroking them with great care and appreciation, the way an instrument maker might handle fine wood. Boxes of ostrich, rhea, pheasant, Wild Turkey, chicken, and goose feathers lined one wall of the wardrobe shop. They kept an array of sizes and colors on hand at all times, ready for the constant repairs required to keep the costumes looking fresh. Many of the pieces date from the show's debut, and replacing them now would be exorbitant—tens of thousands of dollars each for the largest ensembles.

"If you can't afford to do it right, you shouldn't use feathers at all," fashion designer Pete Menefee told me. He and Bob Mackie created all of the show's costumes, and their original drawings guide every repair and modification, living in the shop in large three-ringed binders known collectively as "the bible." "Feathers can last a long time, but maintenance is critical," he went on. "I've seen shows where the green feathers look like spinach hanging off the girls!"

The largest costumes all have nicknames. A fountain of red and orange plumes is known as "Mount Vesuvius"; the green version is called "Asparagus Hat." There are others called "Mohawk," "Crest," and "Smurf." With a fully feathered headdress, backpack, and "butt piece," a dancer's costume can weigh more than thirty-five pounds. It's a lot to balance and carry onstage, let alone perform in. One of the girls explained the dilemma this way: "Not enough swish, and no one looks at you. Too much swish, and your costume falls off!"

The challenge is much the same for male birds of paradise or other strongly dimorphic species. Invest too little energy in plumage, and no one looks at you, but invest too much, and you risk a worse fate: being too awkward or weak to compete, to feed yourself, or to escape predation. The result is a fine balance where slight differences in plumage quality or dancing ability stand out to potential mates, maintaining a constant evolutionary pressure for elaborate feathers.

Here the analogy falters, since showgirls can obviously ply their feminine wiles in contexts other than the Jubilee! chorus line. Nor are their costumes and dance numbers a conscious imitation of any particular bird. "That would look dorky," Pete Menefee said simply. "A real cliché." But the dancers do stand at the receiving end of a long cultural tradition of using bird feathers for courtship and seduction, from the ceremonies at a New Guinean sing-sing, to the subtle accents of nineteenth-century fashion to the bawdy eroticism of a full-fledged cancan or cabaret. And Menefee did have one piece of advice for dancers who wanted to maintain their individuality onstage: red lipstick. "Listen, if you've got thirty-five pounds of feathers on your head and you're not wearing red lipstick, you might as well not even have a face!"

It would be easy to dismiss Jubilee! as little more than an extravagance, the feathered embodiment of Vegas excess. Tourism marketers bill it as the best of "classic" Vegas, a throwback to a bygone era when visitors rubbed elbows with Hollywood celebrities at the Flamingo and other early casinos. But the historical connection runs deeper than Bugsy Siegel. For all of its elaborate flourishes, the show echoes a time when feathers were not just a fashion statement; they embodied fashion. The race to adorn ladies' hats once spawned a global trade where lavish fortunes were made and lost, where bird species were driven to the brink of extinction, and where international intrigues played out like a cross between John Le Carré and Gerald Durrell.

CHAPTER ELEVEN

A Feather in Her Cap

Feathers are suitable for all seasons, as they are always attractive; and since they are of animal fiber and designed by nature to stand all kinds of weather, they usually wear well.

—Charlotte Rankin Aiken,
The Millinery Department (1918)

The slender, soft fingers of women, that handle their own babies so gently and so tenderly, are as hard and cruel as mailed fists when they are stretched out to snatch the wings and plumes and breasts of helpless wild things.

—William T. Hornaday,
New York Times (1912)

In one of Hollywood's most famous scenes, young adventurer Jack Dawson sketches his lover, Rose, reclining on a divan, wearing nothing but a large, blue diamond on a chain around her neck. Their location, of course, is a first-class stateroom on the *Titanic*, mere hours before its sinking. The movie became a blockbuster, entwining their romance with the fate of this fictional treasure, a priceless jewel known as "The Heart of the Ocean."

In reality, the *Titanic*'s doomed hold contained no troves of gemstones, gold, or other obvious riches. The cargo manifest survived and lists items ranging from the mundane (1,963 bags of potatoes), to the odd (28 bags of sticks), to the bizarre (76 cases of dragon's blood—a type of plant resin). There were sardines, mushrooms, lace collars, a case of toothpaste, orchids, and an inordinate number of shelled walnuts. But the most valuable class of merchandise aboard, one that warranted insurance claims worth more than $2.3 million in today's currency, was feathers. The ship contained more than 40 cases of fine plumes bound for the milliners' shops of New York City, and in the spring of 1912 feathers ranked as one of the highest-priced commodities in the world. By weight, only diamonds were more valuable.

The global plume trade peaked in the years before World War I, reaching a scale that seems unimaginable today. Feather merchants and processors in London alone employed more than twenty-two thousand full-time workers, while major hubs thrived in Paris, New York, and other fashionable cities of the day. Though they were used in fans, dusters, boas, floral arrangements, and in the fringes of cloaks and shawls, one fashion craze drove the entire industry: hats. Women did not simply favor feathered hats, they required them. Leaving the house with a bare head was unthinkable, and any respectable wardrobe featured a range of plumed caps and bonnets to fit each season and mood. The most valuable feathers became family heirlooms, transferred repeatedly from hat to hat to keep up with the latest trends, and passed from mother to daughter for generations. After all, feathers had epitomized high style for nearly a half century, and no one could imagine an end to it.

Wild species, known in the trade as "fancy feathers," came into fashion each spring and summer, but one plume remained in style

At the height of the plume boom, feathered hats and accessories featured prominently on the covers of *McCall's* and other popular fashion magazines.

year-round: ostrich. And one country dominated the ostrich business like no other. South African ostrich ranchers once kept more than a million birds in domestication, harvesting their feathers as often as twice a year. Ostrich plumes vied with wool as the country's third largest export, behind only gold and diamonds. Great fortunes were made, and to this day opulent "feather mansions" remain as symbols of excess, their builders as notorious in South Africa as early oil tycoons in America, rubber magnates in the Amazon, or the spendthrift maharajas of colonial India. Ostrich farmers had such political clout that in 1911 they convinced their government to sponsor a secret mission designed to ensure South African dominance in the industry. No other incident exemplifies the money and power at stake in the global feather boom as well as the great Trans-Saharan Ostrich Expedition.

At the very moment the ill-fated *Titanic* struck that infamous iceberg, Russel William Thornton lay near death in the tiny mission hospital at Zungeru, in northern Nigeria. Weakened by severe sunstroke, he'd been bodily carried from his encampment nearby on doctor's orders, leaving the fate of his expedition in the hands of two junior colleagues. A group photograph taken weeks later shows him still emaciated and pale from the ordeal, and for the rest of his life he would never go out in the sun without a hat. But his journal entries record the experience with characteristic understatement: "Nerves bad," "Feeling sick," or, simply, "Ill."

Thornton's bout of sunstroke nearly deprived him of seeing the successful conclusion to one of history's most unlikely adventures. Ten months earlier, the South African government had tapped him to lead an expedition in search of a near-mythic bird, the "Barbary Ostrich." Impressed by his distinguished service in the Boer War, as well as his deep knowledge of ostrich farming, they tasked Thornton with finding, capturing, and bringing home alive as many of the birds as sheer logistics would allow. Money was no object. To a powerful group of lawmakers and feather moguls, this mission was seen as no less than the salvation of the South African ostrich industry.

Known chiefly by reputation and rumor, the Barbary Ostrich boasted extravagant "double-floss" plumes, an exquisitely dense and lustrous feather with more barbs per inch than any other known type. With growing competition in the global marketplace, and with sharp-eyed merchants dividing ostrich plumes into more than sixty distinct grades, such quality mattered a great deal. Cross-breeding these legendary birds with the sturdy Cape Ostrich could ensure South African preeminence in the feather trade for generations. Without such a boost, they felt increasingly vulnerable to the "threat from America," where budding ostrich ranches in Arizona and neighboring states benefited from an ideal climate, irrigated pasturage, and generous government subsidies. Ostrich advocates in America included an influential congressman, who promoted the industry through impassioned speeches to his colleagues in the House of Representatives, filled with statements like "Whoever wears an ostrich plume is adorned with an emblem of justice" or "No one need have any fear for the future of the ostrich industry. The feather is undoubtedly the most beautiful ornament of its kind, and as such is independent of fashion."

South Africa's scheme to counter the American threat with Barbary birds was a great strategy. The only trouble was, no one knew where to find them. The "Barbary states" referred to a vast part of northern Africa, stretching from Morocco all the way to Sudan. For years, government researchers had requested feather samples from agents throughout the region, all to no avail. Though Barbary plumes continued to appear occasionally (and command high prices) in London and Paris, no one could identify the source. The only solid evidence came finally from Tripoli, on the Mediterranean coast, where a parcel of the wondrous plumes arrived with a camel caravan known to have originated in the Sahel, at the southern edge of the great desert. Based on that scant geographical tip, and with the future of an industry possibly hanging in the balance, the Trans-Saharan Ostrich Expedition set forth.

Joined by Frank C. Smith and Jack Bowker, two well-regarded ostrich experts, Thornton departed Cape Town secretly and in great haste in early August 1911. He feared that a rival American expedition might already be under way, led by none other than his own brother. Employed until recently as a government ostrich expert in the Transvaal, Ernest Thornton had abruptly quit his post and departed for the United States under mysterious circumstances. Russel knew that Ernest was fully versed in details of the Barbary plan and suspected that he had "gone over" to the American side. Only later would he learn the truth. Ernest had indeed been courted by American interests but was acting as a sort of ostrich double agent, gathering information on the status of the American industry while letting rumors of his activities spur the South African government into action. As soon as he heard that Russel's expedition had finally been approved, Ernest returned to South Africa, explained his motives in several newspaper articles, and took up ranching again at his farm in the heart of the ostrich district.

Meanwhile, Russel and his colleagues were wending their way by river, rail, and foot more than five hundred miles inland into the vast desert interior of British Nigeria. After logistical stops in London, Lagos, and Zaria, their party had grown to include a guide, translators, headmen, and more than a hundred local porters carrying food, water, baggage, and equipment. They organized ostrich export permits with the regional governor, met several local emirs, and set up a long-term encampment on the outskirts of Kano, a trade center near the border with French Soudan (now Niger). From Kano they organized scouting trips west to the town of Katsina and east as far as Lake Chad and also monitored camel caravans passing through from all directions. Nearly every ostrich feather they encountered proved to be "the wrong type," but finally a pattern began to emerge. The prized double-floss plumes, and the birds that bore them, all originated from villages near a place called Zinder. But although they had

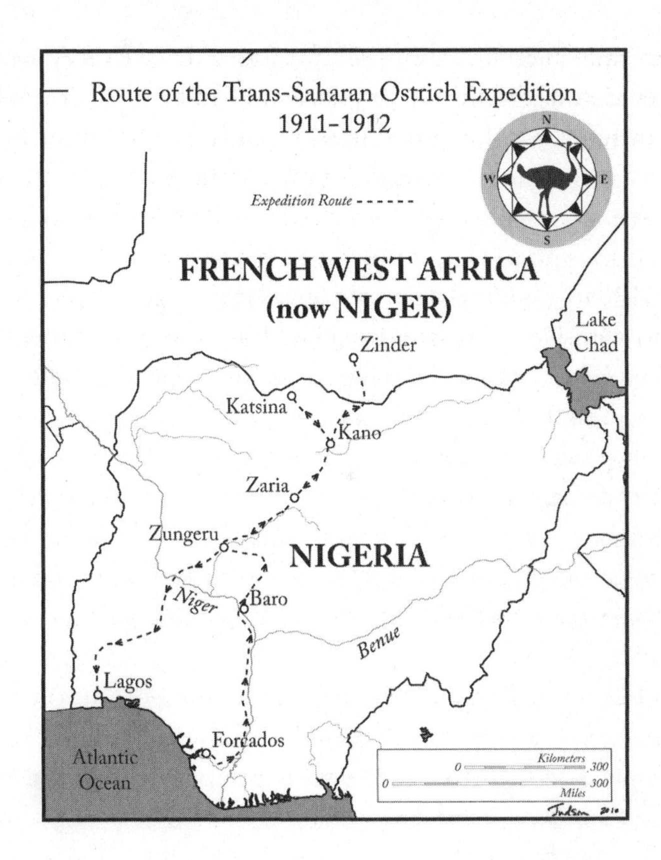

succeeded at last in locating the home of the Barbary ostrich, their triumph was bittersweet. Zinder, and all the land around it, lay deep within French territory.

Not to be deterred, Thornton cabled Pretoria and asked for permission to proceed north across the frontier in pursuit of his birds. Though the government must have anticipated this possibility, they still hesitated for six weeks before giving a final reply. Politically, it was a delicate situation. The Union of South Africa was brand-new on the world stage, formed less than a year earlier from the fusion of Britain's Cape and Natal colonies and the former Boer republics Transvaal and the Orange Free State. While technically self-governing, the union remained part of the British Empire and certainly didn't need a conflict with France to complicate its fledgling foreign policy. In the end, however, the urgency of the ostrich lobby prevailed. Thornton received his permission and all the funding he asked for.

With a journal entry of "Northward Ho!" the expedition set out, reaching Zinder after more than a week of desert treks as long as twenty-four miles per day. Once there, they dined with the commander of the local Legionnaire post and applied for export permits, but the French colonial authorities regarded their objectives with justifiable suspicion. They may not have fully appreciated the value of their ostriches, but they knew they didn't want the South Africans to take them. Though a flurry of international cables brought appeals from the South African and Nigerian governments, and even the British Crown, no amount of diplomacy could alter the governor-general's position. Thornton and his men were not to purchase, pursue, capture, or otherwise procure French territorial ostriches under any circumstances.

Though forced to return to Kano empty-handed, Thornton and his colleagues persisted in their quest and remained in the region for another five months. At this point, however, accounts of the story differ. In his journal and official report of the expedition, Thornton describes slowly acquiring a flock of Barbary birds through his trade connections with local emirs. Though these ostriches surely originated in French Soudan, the South Africans broke no laws and were not directly involved in any cross-border rustling, smuggling, or other skullduggery. Frank Smith, on the other hand, recalled a different version of events.

Later in life, he told wild tales about gun battles with Tuareg raiders, outwitting American spies, and wrangling ostriches from Timbuktu (hundreds of miles from the expedition's actual route).

Though he reportedly apologized to Thornton and Bowker for getting carried away with his public storytelling, his private correspondence suggests that a few of the expedition's exploits might not have made it into the official report: "I am sorry I cannot give the full particulars of how we captured the ostriches in the French Territory and got them across the border into British Nigeria, as this would . . . lead to 'international complications' and must be left out. Between ourselves I may say that we had some quite exciting times outwitting the French Foreign Legion which had been posted along the border to stop our smuggling them across. This however is 'off the record' and definitely *not* for publication."

Just how the birds made their way from Zinder to Kano may be impossible to know, but regardless of whether they were purchased from an emir or smuggled under cover of darkness, the story of the Trans-Saharan Ostrich Expedition needs no exaggeration to make its point. In those heady days at the height of the plume trade, the young government of South Africa was willing to go to extremes for its ostrich farmers. The promise of high-quality feathers merited risking a diplomatic row with France and funding a lengthy international mission in what can fairly be described as ostrich espionage. From Pretoria to Paris to the halls of the U.S. Congress, leaders of the day took pains to protect their interests in the feather economy.

By late April 1912, a flock of 150 "double-floss" Barbary Ostriches crowded the pens around the expedition's Kano encampment. But the challenge was far from over. As mountain climbers are known to tell one another upon reaching the summit of a difficult peak, "Congratulations—you're halfway home!" Thornton lay stricken with fever, and Bowker was in Lagos arranging for a ship, leaving Smith to transport their ungainly cargo across the Sahel and south to the coast. Surviving photographs show the birds being marched along by local porters in flimsy palm-cane pens eight

Photographs of the Trans-Saharan Ostrich Expedition. *Clockwise from upper left:* the caravan to Zinder, ostriches plucked by local feather traders, transporting ostriches by pen, and loading ostriches onto the boat in Lagos.

feet wide. Eventually, they were loaded onto specially designed boxcars, railed to the coast, and then transferred into the customized hold of a steamer bound for Cape Town. Thornton arrived by hammock and rail, Bowker hobbled aboard (he had injured his back in a fall while outfitting the ship), Smith finalized the loading, and they set sail at last.

During their long absence, rumors of the mission had caught the imagination of the South African public, and the men arrived home to the dockside frenzy of a hero's welcome. Amazingly, 127 birds survived the long journey. They were quickly shipped off to a breeding program at Grootfontein, the country's top agricultural college, and for a brief time the future of the industry seemed ensured. No one celebrating on that Cape Town pier could have predicted that in less than two years, the global feather market would utterly implode. Like Cinderella's regalia at the stroke of midnight, Thornton's hard-won Barbary Ostriches changed instantly from priceless to common, from invaluable commodities to a simple flock of birds.

The collapse of the plume boom coincided with the outbreak of World War I and a fundamental shift in women's fashion. Across Europe and America, the war effort brought more and more women into the workforce, and tastes shifted overnight to simpler, more practical attire. At the same time, growing concern for the fate of wild birds was leading to increased legal restrictions on the fancy-feather trade. Demand for plumes of all kinds plummeted, and thousands of feather merchants, ostrich ranchers, and milliners went bankrupt. Some committed suicide. South Africa's stock of more than one million domesticated ostriches quickly dwindled to fewer than nine thousand. As late as 1925, Frank Smith (by then the world's only tenured lecturer in ostrich husbandry) was still lobbying to bring the industry back. As the "special ostrich commissioner" to the 1924–1925 British Empire Exhibition in London, he oversaw a popular display of twenty-four live birds and demonstrated plume-clipping techniques for the queen herself. But though he initially reported seeing plenty of feathers in the fashionable shops on Bond Street, he left London feeling discouraged: "During the two years I was in England I did all I possibly could to try and revive an interest in feathers but it was a hopeless fight—the ladies simply would not have feathers at any price. Small hats and short skirts had come in, and the ostrich feather simply will not 'go' with either of them." The age of plumed hats as emblems of female allurement was over, and Thornton's Barbary flock languished at Grootfontein, breeding experiments unfinished and now irrelevant. By the 1930s they had all died out.

From an evolutionary standpoint, it's worth asking what made the Barbary ostrich so special. Why did this one strain develop feathers of such distinction and quality? As the largest and heaviest of all living birds, ostriches long ago gave up the ability to fly in favor of a fleet-footed terrestrial lifestyle. This allowed their huge wing feathers to adapt principally for courtship displays, an elaborate groveling performance where the sable-bodied males flop down and shake their snowy wings and tail before the tawny females, all the while flailing their long pink necks back and forth, whacking their heads against their sides. Just like in birds of paradise, or for that matter fashionable hats, sexual selection in ostriches favors flamboyance over function. Their flight feathers have lost the asymmetry and interlocking vanes that keep most avian primaries stiff and aerodynamic. The barbs of an ostrich plume fall from the rachis in long, extravagant waves that cluster and droop at the tip like a billow of steam collapsing inward in still air. The double-floss Barbary plume boasted exceptionally dense, lustrous, and luxuriant barbs, an evolutionary quirk apparently restricted to one small population.

Historically, ostriches ranged throughout the arid plains, thorn scrub, and semidesert regions of Africa and the Middle East. Several subspecies and regional variants showed differences in traits like neck color, size, or the thickness of their eggshell. Thornton noted how the tiny range of the Barbary birds was hemmed in on three sides by inhospitable Saharan sands, a degree of isolation that apparently allowed significant genetic distinctions to take hold. The precise answer will unfortunately never be known. Ostriches have been extinct in Niger and Nigeria for more than twenty years.

Habitat loss and human population growth explain most of the recent declines in ostrich numbers. The plume trade played a role early on, but by the mid-nineteenth century, the vast majority of commercial ostrich feathers came from domesticated birds. Ostriches, however, constituted only one facet of the feather market. Wild-caught or "fancy" feathers rode the same wave of popularity, and in some ways their story is even more bizarre.

Like the ostrich trade, fancy feathers reached their peak around the turn of the twentieth century, when virtually any bird that could physically be fixed on a hat became fair game. The offerings at a typical milliner's shop might include individual feathers, fanlike arrangements called "aigrettes," wings, or even whole birds from scores of species, particularly in spring and summer when fashion demanded color and variety. In one famous story, a young banker and avian enthusiast named Frank Chapman went bird-watching on the streets of New York in 1886. He quickly checked off an impressive list of species, but they weren't flying overhead, perched in trees, or pecking at crumbs on the sidewalk. These birds and their feathers festooned the many hundreds of ladies' hats he counted while making his way through a busy market district. Of the seven hundred bonnets, caps, cloches, and downbrims he noted in one of these forays, more than three-quarters bore feathers, and those that didn't were mostly worn by "ladies in mourning" or "elderly ladies," for whom decorum dictated a more muted fashion. From grebes, flycatchers, and woodpeckers to a Northern Saw Whet Owl, Chapman's list included more than forty distinct species, while many more had been mutilated or altered beyond recognition. And he was counting only native birds, those he might have seen if he had ventured into the woodlands of nearby Central Park. Had he broadened the criteria of his search, he could have documented species from almost anywhere in the world: birds of paradise from New Guinea, hummingbirds from Trinidad, Australian parrots, Brazilian motmots, Falkland terns—the growing reach of empires and trade had turned the city's streets into an exotic aviary.

One particular group of birds suffered near extermination at the hands of feather hunters, and their plight helped awaken a conservation ethic that still resonates in the modern environmental movement. With striking white plumes and crowded, conspicuous nesting colonies, Great Egrets and Snowy Egrets faced an unfortunate double jeopardy: their feathers fetched a high price, and their breeding habits made them an easy mark. To make matters worse, both sexes bore the fancy plumage, so hunters didn't just target the males; they decimated entire rookeries. At the peak of the trade, an ounce of egret plumes fetched the modern equivalent of two thousand dollars, and successful hunters could net a cool hundred grand in a single season. But every ounce of breeding plumes represented six dead adults, and each slain pair left behind three to five starving nestlings. Millions of birds died, and by the turn of the century this once common species survived only in the deep Everglades and other remote wetlands. Graphic stories and photographs of the slaughter fueled a small but growing wave of antiplume sentiment, and onetime banker Frank Chapman was right in the thick of it.

Two years after publishing his feathered-hat survey in a letter to the editor of *Forest and Stream*, Chapman left the world of finance behind forever and accepted a lowly post at the American Museum of Natural History. He remained there for more than five decades, rising to eventually chair the Department of Birds and amassing a list of research, education, and conservation accomplishments that led colleagues to dub him "the dean of American ornithologists." From the beginning, his career reflected the same qualities he displayed while observing ladies' hats: creative scientific ideas, a keen eye for detail, and a steadfast sense of outrage at the mistreatment of birds. Though immersed in a range of research and museum duties, Chapman remained an outspoken critic of the plume trade. This passage from his autobiography, describing a visit to a still-unspoiled egret rookery, leaves little doubt about his feelings: "For a time, I was content to sit quietly in the boat and revel in the charm and beauty of the place, my enjoyment unmarred by the thought that at any moment Satan, in the guise of a plume-hunter, might enter this Eden."

As his professional prominence rose, Chapman began actively helping guide the nascent bird-conservation movement. He founded and edited the magazine *Bird-Lore*, which gave a collective voice to the Audubon Society chapters springing up around the country. The name was later changed simply to *Audubon*, and it remains one of the most widely read natural history publications in the world. In 1900, Chapman pioneered the Christmas Bird Count as a nonlethal alternative to traditional holiday hunting parties. The count continues, now boasting more than fifty-five thousand annual participants who document upwards of sixty-five million birds every year in seventeen countries and Antarctica. He proposed the first National Wildlife Refuge to protect breeding birds at Florida's Pelican Island, and President Theodore Roosevelt signed it into law in 1903.

These activities, and the legislative successes that followed, all coalesced around opposition to the fancy-feather trade. Though wildlife of all kinds was under siege at the time, the conservation community found its first real rallying cry in the spectacle of taxidermic birdlife adorning ladies' hats. When Chapman addressed the inaugural meeting of Washington, D.C.'s Audubon chapter, he gave a speech titled "Woman as Bird Enemy." The director of the New York Zoo, William Hornaday, later echoed this theme in his influential 1912 essay, "Woman, the Juggernaut of the Bird World." It began with a gloomy play on words: "The blood of uncounted millions of slaughtered birds is upon the heads of the women."

Ironically, while women made up the feather industry's principal market, they also proved its undoing. Nearly every local Audubon chapter in the nation was founded by women, and they made up the vast bulk of the early membership. Through countless lectures, teas, luncheons, and protests, Audubon activists mounted one of the first grassroots environmental campaigns and made bird preservation a national and international issue. The Lacey Act passed Congress in 1900, restricting interstate transport of wild fowl and game. In 1911 New York State outlawed the sale of all native birds and their feathers, and other states soon followed suit. Passage of the Weeks-McLean Act (1913) and the Migratory Bird Act (1918) took the protections nationwide and mirrored legislation in Canada, Britain, and Europe, effectively ending the fancy-feather era. While ostrich and other domestic feather traders tried desperately to distance themselves from plume hunting's growing stigma, the decimation of wild bird populations left a moral stain on the entire industry and surely added to the change in fashion tastes. In the Everglades, at Pelican Island, and at hundreds of other breeding sites throughout their former range, egret populations began a slow recovery. The Great Egret still adorns the logo of the National Audubon Society. It is pictured in flight, wings tilted and long legs dangling, its breeding plumes streaming out behind like fine brushstrokes from a calligrapher's pen.

On a recent trip to New York City, I took the opportunity to reenact Frank Chapman's famous survey. Though he began on Fourteenth Street and I started on the Upper West Side, I like to think that I covered at least some of the same ground he did. From my small hotel, I wandered down Broadway and passed along various side streets before stopping eventually on the front steps of his longtime haunt, the American Museum of Natural History. The museum building and many of the neighborhood's old brownstones dated to Chapman's era, but a lot of other things had changed in the intervening century. There were still plenty of women wearing hats, and it didn't take long to tally hundreds of them—stocking caps, berets, a few cloches, and even a pillbox or two. (After careful consideration, I decided that sweatshirt "hoodies" didn't count.) Hat *adornments*, on the other hand, were few and far between. I saw several ribbons, some pins, sports logos, and even a pair of long, fuzzy rabbit ears, but there wasn't a single plume in sight anywhere. To find feathered hats in today's Manhattan, you need to make an appointment.

"I'm probably the craziest, most obsessed feather person you'll ever meet," Leah Chalfen told me, moments after unlocking the door to her showroom. Leah C. Couture Millinery occupies a second-floor studio in Midtown, just a few blocks south of the old millinery district. Nowadays, the local storefronts mostly hawk discount handbags and furs, but these same streets once teemed with hatmakers, plume traders, and feather workers bustling among the neighborhood's scores of workshops and factories. Among the tiny handful of people in New York who still make their living crafting hats, only Leah truly specializes in feathers. "I'm doing my best to bring them back," she said emphatically, "even if it's just in a small way!"

Given the results of my hat survey, this seemed like a long shot. But if anyone could pull it off, Leah might just be that person. Petite and fiery, she was enthusiastic enough to make me wonder how feathers might fit into a field biologist's wardrobe. And she had an infectious flair for the dramatic. Throughout the time we talked, she periodically took a new hat down from the wall, tried it on, and struck a pose—all without any discernible pause in the conversation. My trusty blue stocking cap began to seem decidedly dull.

Over the past decade Leah had noticed a small but steady revival in feathers, and she showed me stacks of magazines from around the world that featured her creations. Paradoxically, however, the increase in business had been accompanied by the loss of the last remaining plume wholesalers in the city. The renewed interest, she explained, was all at the top end: custom, high-fashion pieces. "It's a niche market now," she said. "There's just not the volume to support the old supply chain." Then she smiled ruefully and added, "If I didn't make them, I couldn't possibly afford my own hats!"

Leah learned her craft in part by studying vintage featherwork, examining every detail to pick up the old tricks of the trade: how vanes could be clipped and shaped, how a rachis was split, or how to replace the shaft with flexible wire, carefully wrapped in colored thread. "There are no teachers for this stuff. We lost a generation." I asked her why she bothered, what feathers offered that she couldn't get from some other material. "Natural grace," she replied immediately. "There is nothing else like a feather. It's . . . " She paused then, looking for the right word. "It's a glamour thing."

Chalfen's studio filled a high-ceilinged, narrow room that stretched away from the door to a bank of tall windows. It had a pleasing sense of clutter—not the clutter of disorganization but one of creative activity briefly interrupted. For Leah, the space served triple duty as workshop, office, and showroom, with one full wall devoted to the finished product. "Well, there you go," she said as I walked in, gesturing up at the hats with a flourish. "Have at it!"

Leah's signature hat collection is called "The Aviary," and it lives up to its name. Here were aigrettes, colored plumes, and painstakingly crafted wings and "fantasy birds," stylishly arranged on hand-felted hats and headpieces of all kinds. "Don't worry, I use 'barnyard feathers,'" she explained, and pointed to the opposite wall where stacks of boxes held a range of domestic plumes—from rooster hackle and turkey tails to Guinea fowl and ostrich. It was like a miniature version of the Jubilee! costume shop, and Leah's work featured the same incredible attention to detail. But where a showgirl's costume is designed to overwhelm the senses with flash and color, these hats could be delicate, subtle, suggestive. Simply put, the featherwork was exquisite, more like fine sculpture than clothing.

"The most striking thing to me is how they change your silhouette," Leah observed, trying on a soft cap adorned with long copper-striped pheasant plumes. I'd never thought of hats that way, but she was right. The addition of a feathered headpiece instantly reshapes the part of the body most associated with individual identity, the head. Darwin himself noted how the head was the "chief seat of decoration" for birds and people alike. After all, it's where you look first when you meet someone—what better location to adorn for allurement? The cosmetics industry understands this, which is why you find hundreds of varieties of eyeliner, rouge, and lipstick, but no shin liner, rib shadow, or elbow gloss. Unlike makeup, however, which communicates only at close range, a plumed hat can be seen at a distance, making a statement and a first impression before the face even comes into focus.

But shape is only half the story. Milliners follow a long tradition of artists attracted to feathers not only for their structure but also for their incredible palette of colors. Leah's hats ranged from cobalt to burgundy to satiny black, taking advantage of the natural hues and occasional iridescence present in the plumes. Some of her feathers had been dyed, but even "barnyard" varieties offer a wide range of inherent shades. In the wild, birds display every color of the rainbow and beyond. What's more, their eyes perceive a much wider spectrum than ours, so hats or plumage we think of as vivid must shout out to birds with a vibrancy we cannot even imagine.

Feathers in modern fashion—a hat from Leah C. Couture Millinery.

Leah and I stayed in touch after the interview, our disparate worlds brought closer by a shared feather obsession. When I sent her pictures of a King Bird of Paradise displaying its cascade of crimson plumes, she responded, "INCREDIBLE—blows your mind!" It seems that no matter how long one works with them, or how much one learns, feathers still retain the power to amaze, to inspire wonder. The sheer brilliance of bird plumage, and our visceral reaction to it, begged a host of new questions. After talking with Leah, I realized that my exploration of fancy needed to delve beyond the history of plumes, bird hunters, and hats and into the evolution of color itself.

CHAPTER TWELVE

Give Us Those Nice Bright Colors

They give us those nice bright colors
They give us the greens of summers
Makes you think all the world's a sunny day, oh yeah
I got a Nikon camera
I love to take a photograph
So Mama don't take my Kodachrome away.

—Paul Simon, "Kodachrome," 1973

"We can't show you the process," he said flatly. "We can't even talk about how we dye the feathers." The voice on the line sounded gruff and wary, like someone used to fending off the curious. I assured him that my interest was purely academic, that I just wanted to stop by for a chat about the feather business. He sounded doubtful, but with a little persistence, he finally put me through to the owner.

"We can't show you how we dye the feathers," she told me immediately, and for a moment I thought the conversation was over.

But she didn't hang up, so I started explaining about the book project, and pretty soon she got curious. For people in the small world of feathers, there's a sense of shared fixation that usually trumps everything else, even when carefully guarded trade secrets are at stake. We spoke for only a few minutes before she invited me down to the shop for a visit.

The Rainbow Feather Company occupies a low cinder-block building five blocks and a million miles from the Las Vegas Strip. The neighborhood is light industrial, home to discount tire stores, an auto body shop, metal fabricators, the Opportunity Village Thrift House, and Acme Bail Bonds. On the day I walked through, a dry desert wind whipped grit into the air and plastered trash bags to the chain-link fences circling vacant lots and car parks. It was like going backstage to see the mundane pulleys, ropes, and winches that made Jubilee!'s elaborate props and dancers soar, a glimpse of the everyday, nuts-and-bolts businesses that prop up all the Vegas glitz.

In that sense, Rainbow Feather was right at home. Though it seems a long step removed from the stage and bright lights of a show like Jubilee! the work done in that nondescript building was every bit as important as the choreography, the music, and the showgirls. The glamour of a Vegas show depends on costumes, and the costumes depend on feathers, and the feathers must be bright and colorful. And when you need ten thousand plumes died hot pink, orange, yellow, green, or any other color, Jodi Favazzo is the only person in North America who can help you.

"I've been around feathers all my life," she told me, explaining how her mother did piecework from home, crafting little feathered flowers for the hat and craft industry. Frustrated by the lack of colors available at the time, she convinced her husband to start dying small batches for her in the kitchen sink. Soon he'd left his construction job to found the Rainbow Feather Company as a full-time enterprise. That was nearly fifty years ago, and it's been a family operation ever since.

"You can't dye feathers with a machine," Jodi said. "It has to be done by hand." We were inspecting an array of her finished products in the small retail space at the front of the building. It was set up like a clothing store, but instead of jeans and sweaters, the racks held long strands of brightly colored turkey, goose, duck, and chicken feathers, as well as ostrich boas of every imaginable shade. There were whole bird skins, too, and packets of loose plumes, not to mention bins bristling with peacock and pheasant tails. For their custom work, Rainbow Feather takes orders from clients as varied as Jubilee!, Cirque de Soleil, and Victoria's Secret, but the retail shop struck me as even more intriguing. It must be the only place in the world where burlesque dancers regularly rub elbows with fly-fishermen and bow hunters. "We sell to everyone," Jodi confirmed, and while we were talking, a man came in to pick up a headdress for his Mardi Gras costume.

A trim, attractive woman with perfect posture, Jodi looked like she could have had a career onstage if she'd cared to. But life behind the scenes suited her just fine. "This is a great business," she told me at one point. "I come in here every day, and I'm amazed."

My visit had interrupted her in the middle of a batch, and I could still see traces of turquoise dye on her hands when she held up some rooster hackle and started explaining the details. "From the time they're plucked, all the feathers from each side of the bird have to be kept separate. If they're mixed they won't string right, and they rub against one another in the vat and come out streaked." She told me that no two feathers took a dye in quite the same way. "It depends on the bird's environment—what it was eating, the weather, the minerals in its water. All of that affects the feather."

Dying a feather is a surprisingly tricky, multiple-step process. The first challenge lies in bleaching out the natural pigments and preparing the keratin to soak up new colors. "In the old days my father used sulfuric acid and wore rubber boots, gloves, the full suit," she explained. The process is more benign now, using a combination of chemicals known only to Jodi, her husband, and a few other family members. When all goes well, the feathers come out white and supple, ready to take on any shade that Jodi dreams up. People often send in color swatches for her to match, just like ordering paints at the hardware store.

"We can do any color, but there's one thing that sometimes frustrates people," she said, and led me over to the Wild Turkey flats and pheasant tails. Their barred vanes had been dyed to various hues, but they still glowed with their natural bronzy sheen. "We can dye them, but I can't take out the iridescence," she explained. "That's just part of the feather." With that simple observation, Jodi summed up the physics of feather coloration as neatly as any textbook.

As the birds of paradise demonstrated, sexual selection and female choice have played a powerful role in the development of showy displays, but camouflage, social signaling, parent-offspring recognition, and a range of other functions also added pressure for the evolution of color. Over time, feathers responded with two main strategies: pigment-based colors and structural colors. Though the resulting shades sometimes look similar, the two approaches differ in the fundamental way they treat light waves.

Pigments work by selective absorption. When light hits a pigment-bearing feather, part of the spectrum is absorbed, and the rest is reflected back to our eyes as a color. If all the light is reflected, we see white. If it's all absorbed, we see black. Gradations in between give us a wide range of hues, from the earth tones of a Song Sparrow to the yellow on a Goldcrest to the brilliant red crown of a Pileated Woodpecker. Pigment-based colors are familiar to us—the melanin tinting a sparrow feather is the same molecule that darkens human hair and causes skin to tan. We paint our houses and cars with pigments and use them to color our clothes. When Jodi Favazzo dyes a feather, she is stripping out the natural pigments and replacing them with those of her own choosing. The same thing goes on every day at hair salons around the world.

Some pigments are easy for birds to manufacture right in the cells of their developing feathers, while others, particularly yellows and reds, must be acquired through their food. Flamingos, for example, remain pink only so long as their diet includes a healthy dose of algae and crustaceans rich in beta-carotenes (the family of pigments responsible for the red in a lobster shell and the orange of carrots). Captive birds grow paler with every passing molt and will ultimately turn white unless their feed is supplemented with dyes. New eating habits can also change the coloration of birds in the wild. When Cedar Waxwings in the northeastern United States feed on berries from an invasive, nonnative honeysuckle, the tips of their tail feathers shift dramatically from yellow to orange the next time they molt, responding to an unfamiliar pigment in the fruit.

A Cedar Waxwing.

Pigment-based coloration accounts for a wide range of feather patterns and shades, but structural colors generate some of the most dramatic displays. The scarlet iridescence of a hummingbird throat, the metallic glint of motmots, and the radiant blue of a jay—these colors are all created not by absorption but by the scattering of light. For structural colors, the whole spectrum is reflected back from the feather surface by nanoscale features built into the keratin. If that reflection is a random scatter, we perceive it as white, but when the wavelengths are ordered, we see them as rich, shimmering colors.

To understand how physical structure alone can create something so vivid, I decided to conduct an experiment. I did the dishes. In our house we use a biodegradable dish liquid in an old-fashioned ceramic sink. The soap is clear and pigment-free and so is our tap water, so without some kind of structural phenomenon, filling the basin with dishwater should be an entirely colorless experience. I plugged the drain, ran the faucet hot, and squeezed in a good dollop of soap. Predictably, a thick foam quickly bubbled up and spread across the surface of the water. Two structural effects became immediately apparent. Where the bubbles were small, the foam looked as fair as the driven snow—light striking their complex surfaces was scattered in a million directions, a randomness my eyes saw as white. But the larger bubbles shimmered, their taut surfaces twisting the light into iridescent prisms of red, violet, blue, and orange. If I popped them or swept the foam aside, the soapy water looked clear again, and I could see straight through to the bottom of the sink. Without the structure inherent in the bubbles, all color disappeared, but when they were present, I could see rainbows.

In feathers, structure and pigment often go hand in hand. The luminous green of parrots, for example, comes from a blue structural color at the surface blended with underlying pigment-based yellows. Knowing this, I eagerly scooped up a layer of dish foam onto a bright-yellow salad plate and peered at it intently. It did not look like a parrot. The white foam remained white, the plate stayed yellow, and the sheen on the big bubbles still held the same swirling mix of colors. Clearly, there was more to the intricacies of feather coloration than I could learn at the kitchen sink.

Whole careers have in fact been devoted to it, revealing scores of different structural designs and nuances of pigment. There are crystalline lattices and self-organizing matrices, complex metabolic pathways, and molecules stacked like pancakes or carefully arranged around air pockets. Every combination tweaks light in a subtly different way, and the result is an unparalleled diversity of colors and effects. To complicate things further, birds perceive a whole range of ultraviolet shades invisible to the human eye; they see in what one expert calls "a third dimension of color."

Although the physics of feather colors may be complex, their evolutionary history recently got a whole lot clearer. When I met with Rick Prum, he showed me an artist's rendition of *Anchiornis huxleyi*, the feathered dinosaur that Xing Xu had found deep below the Yixian shales. *Anchiornis* first made headlines for predating *Archaeopteryx*, but now it had another claim to fame. By examining the specimen with an electron microscope, Prum and a group of colleagues found evidence of coloration in the presence and arrangement of certain molecules. "We literally have the tools to start writing an illustrated field guide to dinosaurs!" he said excitedly, as if he could hardly believe it himself. And if *Anchiornis* is any indication, it will be a colorful book. Prum's picture showed an animal with dramatic black-and-white-striped plumes and a fiery reddish crest, like a four-winged woodpecker with teeth. The discovery suggests that colorful plumage has been around as long as feathers themselves, strengthening the argument that beauty and display played an early role in feather evolution.

If dinosaur feathers were bright, then birds evolved with color from the very beginning. This long history helps explain how feather pigments, structures, and their resulting palette became

Fossilized pigment molecules reveal that *Anchiornis huxleyi,* the oldest feathered dinosaur discovered to date, sported pied wings and a fiery reddish crest.

so varied and so intimately involved in avian courtship. But birds are not the only animals attracted to color. Even before human hunters co-opted plumes to guide their arrows, human artists gathered them to express their creative visions. Before the invention of modern acrylics, dyes, oils, and pastels, what other medium offered the artisan such brilliance? Fish could be gaudy, but they faded quickly out of the water. Butterfly wings were too fragile, beetle backs too brittle, and gemstones too rare. For hunter-gatherers and even early civilizations, only bird feathers offered ubiquitous, varied, and durable color. Human cultures everywhere co-opted them for artwork and handicrafts, to decorate their bodies, and as symbols of status.

No one can put a date on the first feather decoration, but archaeologists continue to find older and older examples of other bird-derived handiwork. The world's earliest known musical instrument, a flute, was carved forty thousand years ago in Germany from the hollow wing bone of a Griffon Vulture. Excavations near France's famed Lascaux Caves have revealed bird-bone needles, pendants, and beads, as well as bird-bone flasks that the artists used to hold their ochre pigments. Although any feathers at these sites have long since rotted away, it's hard to imagine ancient musicians, painters, and artisans crafting the tools of their trade from the bones of birds without also finding a creative use for their colorful plumage.

In many cultures, the importance of featherwork has lasted into the modern era. As recently as the 1970s, a young man on the South Pacific island of Santa Cruz could only get married by paying a bride price composed exclusively of feathers. And not just any plumes would do. Feather money, or *tevau*, consisted of long, intricate coils crafted from the head, neck, and back feathers of the Cardinal Myzomela, a brilliant scarlet and black honeyeater indigenous to the Solomon Islands. Traditionally used for other large purchases as well (e.g., canoes, pigs, houses), each feather coil measured up to thirty feet (ten meters) in length, required the plumage of 350 to 1,000 birds, and took more than seven hundred hours to construct. Only a few families knew the crafts involved in manufacturing *tevau*, from capturing and plucking the honeyeaters to individually hand-gluing tens of thousands of tiny plumes onto the coils (which were themselves composed of pigeon-feather discs woven together with bark and fiber ropes). Good marriage prospects might fetch ten coils or more, hefty sums paid in drawn-out installments that bound the community together in a web of feather debts and obligations. *Tevau* may be the most elaborate example, but feathers, as well as shells, served as currency throughout the Pacific, where trade was advanced, but the atolls usually lacked any kind of metal ores, gemstones, or other sources of durable color.

Unlike the Santa Cruz islanders, who hoarded their beautiful feather coils in the smoky lofts of their huts to preserve them from

The feather money of Santa Cruz Island was made from the scarlet plumage of Cardinal Myzomelas. *Left:* a hunter's catch. *Right:* finished coils.

insects, most plumed handicrafts were made for display. In Hawaii, chiefs and royalty commissioned elaborately plumed capes and ceremonial helmets, where the color and rarity of the feathers helped establish their status. No color was more scarce in the islands than yellow, and no royal object more famous than the flowing golden cloak of King Kamehameha I. Feathers from an estimated 80,000 individuals of the now extinct Mamo Honeycreeper went into its lush weavings. Traditional featherwork continues in the art and adornment of many cultures, from the Waorani and Karajá nations of the Amazon Basin to the Karo of Ethiopia to the Akha hill people of Thailand, Laos, and Burma. But among all feather artisans in history, perhaps none took the craft to greater heights than the empires of pre-Columbian America.

When Hernán Cortés arrived in the Aztec capital, Tenochtitlán, in 1519, he called the island city "the most beautiful thing in the world." He and his men marveled at the canals, the aqueducts, the temples, and the floating gardens. Of Montezuma's palace he told his own King Charles it was "impossible to describe its excellence and grandeur . . . save to say that in Spain there is nothing to compare to it." Yet perhaps no structures in the city elicited more wonder than the aviaries.

Fashioned like palaces only slightly less grand than Montezuma's own, they held dozens of courtyards, balconies, and gardens; ten artificial pools (both saltwater and fresh); and long corridors of rooms with latticework ceilings open to the air. Three hundred attendants looked after the flocks, including veterinarians and dedicated keepers responsible for bringing in enough chickens, worms, maize, and grain to keep each bird "fed with the food which it eats when wild." The cormorants, herons, and other piscivores alone required more than 250 pounds of fresh fish every day. Cortés described the aviaries holding "every species of bird known in these parts," while one of his soldiers, Bernal Díaz del Castillo, later wrote in more detail:

> I am forced to abstain from enumerating every kind of bird that was there . . . for there was everything from the Royal Eagle and other smaller eagles . . . down to tiny birds of many-colored plumage, also the birds from which they take the rich plumage which they use in their green feather work. . . . [T]hen there were parrots of many different colors, and there are so many of them that I forget their names, not to mention the beautifully marked ducks and other larger ones like them. From all these birds they plucked the feathers when the time was right to do so, and the feathers grew again.

But though these great flocks were carefully bred, cared for, and plucked, they supplied only a fraction of the feathers needed for Aztec artisans. Nobles and wealthy families maintained their own smaller aviaries, and even common citizens kept colorful songbirds as pets. The emperor and his governors also demanded feather tithes from conquered territories and sent feather merchants and hunters abroad to bring back plumes from as far off as modern-day Panama and Colombia. This trade greatly enriched the Aztec color palette, bringing the pinks of coastal spoonbills and the brilliant hues of lowland parrots, as well as the plumes of Keel-billed Toucans, Shining Honeycreepers, Pygmy Kingfishers, and other bright species not found in the high country at the heart of the empire. Ornithologists have even suggested that Aztec and other pre-Columbian bird traffickers permanently altered the ranges of several Central American species, introducing Great-tailed Grackles to the valley around Mexico City and the Tufted Jay to the highlands of western Mexico.

From this wealth of avian color the Aztecs fashioned headdresses, mitres, shields, cloaks, tapestries, and ornaments of all kinds. In folktales, the emperor never wore the same set of clothing twice, passing on his elaborate garb as a reward to favored nobles at court. True or not, Montezuma apparently had a wardrobe large enough to spare the two or three "loads of cloaks of rich featherwork" he lavished on every Spanish soldier traveling with Cortés. Among the feather art given to, or later taken by, Cortés and his men were elaborate images of monsters, serpents, animals, and birds "painted" with shimmering feathers "more wonderful than anything in wax or embroidery."

Sadly, Cortés and later governors followed a doctrine of cultural repression and replacement, outlawing traditional practices, including featherwork, and destroying countless artifacts. Of the great Aztec featherwork, fewer than ten pieces are known to sur-

Aztec feather art ranged from tapestries to Montezuma's royal raiment to the uniforms and shields of the common soldiers shown here.

vive. They include a headdress of Resplendent Quetzal tail feathers in a Spanish museum and a faded coyote shield in Vienna crafted from the plumes of Blue Cotinga, Scarlet Macaw, Yellow Oriole, and Roseate Spoonbill. Montezuma's aviary was burned to the ground during the Spanish siege of Tenochtitlán in 1521, specifically targeted by Cortés to "distress" the enemy. Flames from the latticework and heavy timbers were reportedly visible throughout the city and from all shores of the lake.

To the south, the Incas and their predecessors in Peru pursued a similar trajectory of empire, plume trading, intricate featherwork, and rapid postcolonial decline. They too set up a system of aviaries, with tithes paid in birds and feathers from all corners of their domain. Scores of species were used, many transported up and over the mountains from vassal kingdoms and trading partners in the Amazon Basin. Inca bird handlers even learned to manipulate the natural coloration of feathers. By rubbing the skin of captive parrots with the secretions from poison arrow frogs, they created a whole new palette in the next molt, transforming the birds' normal greens and reds into deep golden yellows and salmon pinks. When the Spanish arrived, they admired the craftsmanship of Inca featherwork just as they had that of the Aztecs. But again they associated it with paganism and potential resistance to colonial rule, prohibiting its manufacture and destroying huge stores of feathered textiles and artwork. Centuries of accumulated skills and tradition quickly disappeared, but many examples of Peruvian featherwork managed to survive. Unlike the Aztec situation, where anything the conquistadores missed quickly succumbed to rot, Peru's artwork benefited from the winning combination of elaborate funeral rituals and an arid climate. Archaeologists continue to uncover beautiful examples of feathered tunics, tabards, helmets, headdresses, statuary, bags, and shields. Sealed in dark, dry tombs scattered across the coastal deserts, many of the plumes have retained their vibrant coloration for more than a thousand years.

Parched desert burial chambers are a perfect place to keep old feathers. Protection from light preserves their colors, while the lack of moisture prevents bacteria and fungi from breaking them down. In fact, climate turns out to be an excellent predictor for the survival of ancient American featherwork: quite a bit from Inca deserts, very little from the temperate Aztec highlands, and nothing from the Mayan rain forests. Museum curators consider a leaky roof as perhaps the worst enemy of feather preservation. I learned this from Marian Kaminitz, head of conservation at the Museum of the American Indian in Washington, D.C.

We were sitting in the museum library, surrounded by plastic tarps covering the books, shelves, carts, and tabletops. Thankfully, the leak hadn't reached the collections warehouse, a cavernous two-story room filled floor to ceiling with artifacts packed into white archival cabinets. Nor had any rain dripped into the lab, where we had just watched a technician making careful repairs to a striking jet-black dancing skirt from Southern California. She cleaned each plume with tiny dabs from a dry sponge and painstakingly reconnected loose barbs in the vanes. The piece was irreplaceable—it dated from the nineteenth century, and every feather in it came from the now critically endangered California Condor.

"We're a living cultural museum," Marian explained, noting how she and her colleagues consult and interact regularly with tribes from throughout the Americas. Although the bulk of the collection is historic, the museum still acquires new pieces from groups whose artistic traditions, including featherwork, continue to flourish. That living connection allows a much deeper understanding of the artwork and why the use of a particular feather might have been significant. Unlike paints or dyes, feathers bring more than just color to a composition. They come imbued with all the characteristics, symbolism, and mythologies associated with their species: the raven as trickster, the wise owl, or hummingbirds as an incarnation of the sun. In some cases, the feather colors themselves are explained by folklore. Reds and yellows came from bathing in sacred blood or fire, blue from the skies and rivers or, to the Cashinahua people of the western Amazon, from the pierced gallbladder of a mythical beast slain by bird warriors.

Understanding such subtleties in featherwork is sometimes impossible without a cultural guide. Marian recalled preparing a set of Yu'pik Eskimo dance fans for an exhibition at the museum.

They were quite old, and the men's fans were missing most of their Snowy Owl feathers. Replacing the plumes went against her historian's impulse to preserve artifacts "as is," but a Yu'pik elder told her the fans would be utterly meaningless otherwise. Minor wear and tear could be ignored, but the owl feathers were integral to the dance itself. "He explained it to me this way," she said. "'If you get a scratch in your car, you can still drive it. But if it doesn't have a carburetor, it won't go.'"

The carburetor analogy is a good reminder that no matter how beautiful a feathered artifact, there was a practical cultural logic to how and why it was created and to how it was used. Birds too show a certain pragmatism in the evolution of their plumes. Although a discussion of coloration inevitably focuses on the showy birds, one need only glance through the nearest field guide to realize that most species are actually quite drab. "Little Brown Jobs" outnumber the Resplendent Quetzals of the world by a large margin, and even in fancy birds, the females and juveniles are usually some shade of brown. Brightness evolves for show, but when showiness isn't needed, it's better to blend in. The frequency of dark and mottled feather patterns makes a good case for camouflage as the most successful color scheme of all.

But whether a bird is brilliant or dull, its feathers have uses beyond the intuitive categories of flight, insulation, and display. The great diversity of feather structures leads to a great diversity of function, both in nature and in the realms of human invention. Our exploration turns now to all new territories—functions underwater and on top of it, in patent applications, on parchment, in musical rain-forest trills, and inside the rotted stomach of a zebra.

Function

Cackle, cackle, Mother Goose,
Have you any feathers loose?
Truly have I, pretty fellow,
Half enough to fill a pillow.
And here are quills, take one or ten,
And make from each, pop-gun or pen.

—Mother Goose traditional rhyme

CHAPTER THIRTEEN

Of Murres and Muddlers

An ingenious angler may walk by the river, and mark what flies fall on the water that day, and . . . if he hit to make his fly right, and have the luck to hit, also, where there is store of Trouts, a dark day, and a right wind, he will catch such store of them as to encourage him to grow more and more in love with the art of fly-making.

—Izaak Walton,
The Compleat Angler (1676)

"Try not to kill anything on your way home!" a friend called out as I left the dinner party, and everyone laughed. The roads were good and I had only a few miles to go, but he wasn't entirely kidding. After all, I was driving the Truck of Death.

On the surface, it looked like a perfectly normal Toyota pickup—gray, with a white stripe and a matching canopy. I'd gotten a pretty good deal on it used, but quickly began to realize why the previous owners had been eager to sell. The engine ran fine and the body was in great shape, but that truck had a strange and disturbing habit of killing any animal that happened to stumble into its path. I often arrived at work with chickadees or juncos smashed in the grill, and once found a kinglet wedged under the arm of the windshield wiper. Then I hit a rabbit. Then two more rabbits, a cat, two crows, a robin, and an indeterminate number of voles and other small mammals. Recently, I had flattened a four-point buck in the middle of a national park.

As a biologist, I'm accustomed to collecting the occasional specimen or two for the sake of science, but before buying that Toyota, the death toll from my entire driving career had amounted to one Rufous-sided Towhee. I don't know if it was the color, the shape, or something more sinister that made that particular pickup so deadly, but I'd begun to dread getting behind the wheel. What made it particularly creepy was that, like something out of a Stephen King novel, the Truck of Death came through every collision without so much as a broken headlamp or a scratch in the chrome.

So when my headlights picked out a small dark lump lying in the roadway ahead, I knew enough to hit the brakes. It looked like a discarded T-shirt or a bundle of rags, but as the truck rattled to a stop, front tires only a few feet away, the rag pile suddenly peered up at me and blinked.

The word *incongruous* comes directly from the Latin for "not fitting" or "out of place." Finding a strictly oceangoing bird sitting placidly in the middle of a dirt road seemed to fit the definition rather nicely. Common Murres are members of the family Alcidae, the auks, a group of sturdy, plump seabirds that use their wings to "fly" underwater in pursuit of prey. They spend their lives entirely at sea, making a brief annual stop at coastal cliffs or rocky islets to breed. It appeared that my pickup truck, having run out of new terrestrial fauna to kill, had found a way to start attacking ocean creatures.

The murre looked perfectly calm and composed sitting there in the headlights' glare, as if I were the one in the wrong habitat. But it was obviously disoriented, having mistaken the flat surface of the roadway for a calm stretch of water. The same confusion occasionally afflicts whole flocks of alcids, who've been known to come careening down onto wet parking lots or airport runways. Once grounded, they can only flop about awkwardly and have little hope of getting airborne again—their plump, heavy bodies require the buoyancy of water and a long running start to take flight. I pulled over and parked, hoping the murre's landing hadn't been too hard. They're tough, well-padded birds, and I wasn't worried about broken bones, but a single damaged feather could be the difference between life and death.

The murre let out a hiss and lurched away as I approached, but this was not muttonbirding, and there was no burrow to retreat to. I ran forward and grabbed it from behind, pinning its wings carefully to its body. At first it struggled, then turned its head and bit me, clamping down hard on the fleshy part of my thumb with its stout, strong bill. This had an immediate calming effect, something I've noticed in other birds and even some small mammals: once they get their fangs into you, they feel their job is done and it's all right to relax and enjoy the ride. The bird held on tight as I turned it upside down to inspect the belly feathers. They looked fine. No scuffs, no broken shafts, no down poking up through the plumage; just a perfect white smoothness reflecting brightly in the headlights. This bird was ready to go back to the water.

At that point I realized that my plan had a few flaws. I was holding the murre, the murre was biting me, and the beach was a mile away. I had no way to put the bird down and nowhere to put it. I couldn't possibly drive—I didn't even have a free hand to turn off my headlights. That's how I found myself walking down a country lane in pitch darkness, talking soothingly to the seabird chewing on my hand. More recently, I've employed similar techniques to get my infant son back to sleep, but on that night it gave me another occasion to marvel at feathers.

Had I found a single blemish in that bird's plumage I would have had to take it home and get it to the local wildlife rehabilitation center, where it would have stayed indefinitely, living on old bait fish and cat food until its next molt. Returning a seabird to the ocean with its feathers in disarray would be a fate as certain as leaving it in front of the Truck of Death. Water draws body heat from unprotected skin at rates up to twenty-five times as fast as air. In the chilly currents off our island, an unprotected person can show signs of hypothermia in ten minutes and rarely survives as long as an hour. Birds, with their tiny body mass, would last a fraction of that time unless they were safely sealed inside their feather coats. From my stranded murre to Emperor Penguins leaping through holes in the Antarctic pack ice to the Mallard cadging bread crumbs in a city park pond, any bird living a life aquatic must be watertight. It's an odd paradox: water birds never get wet.

For generations, ornithologists attributed this phenomenon to preen oils, the waxy secretions that birds spread liberally on their plumage during daily bouts of grooming. Watch any preening bird and you will see it repeatedly twist around to root at a spot directly above its rump, but this behavior has nothing to do with scratching an itch. The bird is simply loading up on oil from its preen gland, a specialized organ whose lipid-rich secretions help keep feathers supple. At first glance, the connection to waterproofing seems obvious: oils are notoriously water-repellent, and the largest known preen glands are found in aquatic birds, species like prions, ducks, and pelicans. Later studies suggested that specialized feathers called powderdown also played a role. Again, the logic seemed intuitive: they shed tiny keratinous flakes that had the drying properties of talcum powder, and many birds with the most prolific powderdown appeared to have reduced or absent preen glands.

Intuitive logic needn't be true, however, so I decided to test it. I took a flight feather from a road-kill goose out on the porch of the Raccoon Shack and poured water on it. The liquid beaded up and ran off immediately in perfect silvery droplets, leaving no trace of dampness behind. Under magnification, I watched the water stream off the rachis and form into jewel-like drops, perching on the intricate barbed lattice of the vane. The underside of the feather remained absolutely dry. But whether this was due to some lingering residue of powderdown or preen oils, or the structure of the feather itself, was no clearer. I needed to step my test up a notch and take some advice from Jan Miner.

For twenty-seven years, Miner, an American actress, starred in a famous series of television commercials, portraying a wisecracking manicurist named Madge. Each installment featured her dunking some unsuspecting woman's fingertips into a bowl of Palmolive dish soap and then nattering on about its softening virtues before revealing, "You're soaking in it." Madge's liquid detergent softened her client's skin and nails the same way it cleaned plates, by breaking down grease and natural oils into tiny particles, allowing water to penetrate to the underlying surface. With this in mind, I took my goose feather to the house and subjected it to a good hot scrubbing in the kitchen sink.

It emerged from the suds looking utterly ruined, wet barbs clinging to the rachis or clumped together in dark mats. Once the feather dried, however, it quickly regained its familiar shape, and even my clumsy attempt at preening managed to rezip most

Seen in cross-section, this scanning electron microscope image shows a water droplet perched on the barbs of a Common Pigeon feather.

of the barbs into a smooth vane. I applied water and once again watched it bead up into pearly droplets. The message was clear: even with its preen oils gone, the goose feather was waterproof.

Most feather researchers now agree that structure is the key to waterproofing. Their experiments have subjected feathers to cleaning by harsh detergents or ethyl alcohol, and they have come up with pressurized contraptions to force water and air through the plumes. In spite of such treatment, flight feathers and contour feathers demonstrated their resistance to water time and time again. Only down feathers appear to be soakable, and even those offer some level of water resistance. Mallard ducklings manage to keep dry within their natal down on regular swims just a few days after hatching, before their preen glands have even started producing oil.

The lightweight, efficient, easily repaired, and extremely effective waterproofing microstructure of feathers is beginning to attract a lot of attention, and not just from ornithologists. Physicists, engineers, and inventors have an interest in water resistance, too, and top-flight research on feather structure now appears regularly in publications such as the *Journal of Colloid and Interface Science* and the *Journal of Applied Polymer Science*—very far from ornithology's stomping grounds. Some of the titles indicate how the rewards have changed: this isn't just about intellectual satisfaction. As the makers of Gore-Tex and other Teflon-derived fabrics have shown, waterproof materials are a multibillion-dollar business. But where Teflon production requires polluting chemicals like perfluorooctanoic acid, feathers are waterproof naturally and may hold the key to an environmentally friendly alternative.

Just how they do it remains something of a mystery. I contacted a Chinese scientist whose team had studied the microroughness of plumes from twenty-nine different bird species. They calculated the number of "touch points," all the tiny edges and corners in the lattice of a feather vane that come into contact with water. Most birds have dozens of touch points in every square millimeter of feather surface, but that number increases dramatically for many water birds, up to an incredible nine hundred distinct points per millimeter in a Jackass Penguin feather. He told me it was the density of touch points that determined water resistance—all those distinct places pushing against the natural surface tension of the water. On the other hand, an Israeli physicist was adamant when he explained to me it was the air pockets trapped between touching points that actually repelled the water, not the points themselves. He had taken electron microscope images of a tiny water droplet perched atop a pigeon feather, with air pockets clearly visible between the feather barbs below. Increased density of barbs (and barbules and barbicels), he argued, would increase the number of air pockets and the degree of water resistance correspondingly.

The mechanics of what actually happens when water touches the surface of a feather remain very much in question, but one thing is certain. Considering their light weight, flexibility, and thinness, feathers offer one of nature's most versatile and efficient waterproofing membranes, a feat of engineering that scientists (and the folks at Gore-Tex) would love to sort out.

In the meantime, biologists are using new insights on feather structure to sort out some long-standing questions about water birds. Cormorants and shags, for example, provide an interesting twist on the waterproofing story. A family of long-necked divers famed as trained fishing birds in Japan and China, they occur around the world and share a distinguishing trait: their outer feathers get soaked every time they plunge below the surface. From the standpoint of diving, this gives them an obvious advantage—less air trapped in their plumes means less buoyancy, making it easier to stay under and chase after the fish and crustaceans that make up their diet. (Loons, another agile diver, have unusually solid, heavy bones for the same reason.) Early observers thought cormorants must lack functional preen oil glands, surviving instead by a combination of dense underplumage and their habit of perching for long periods between dives with their wings extended, apparently air-drying their feathers. But closer examination revealed typical preening habits and glands of normal size and function; once again the real answer lay in feather structure.

A cormorant contour feather is loose and wettable only around the periphery of the vane. Toward the rachis, the barbs become increasingly dense, with nearly as many touching points as a penguin feather. A waterproof layer is maintained by tight overlapping of the plumage, with no gaps between those densely barbed interior vanes. In this context, the shags and cormorants come off looking rather smart. Rather than unkempt survivors with wet feathers and a crummy preen gland, ornithologists now view them as beautifully adapted to a diving lifestyle. They benefit from the negative buoyancy of soaking, while still keeping their skin and down feathers sealed inside a watertight blanket.

While the need for waterproofing is obvious for diving cormorants or a stranded Common Murre returning to its cold sea, every bird species lives exposed to the weather and needs a way to avoid getting drenched. I once watched a Rufous Hummingbird huddled over her nestlings during an unseasonably cold downpour. I felt damp and chilled in a parka and a wool sweater—it seemed impossible that such a tiny bird or her chicks could survive. But of course the water simply sluiced off the feathers on her back and wings, keeping everything dry below. Later that spring, all the youngsters fledged.

For birds, the outermost layer of plumage is a vital barrier to the elements no matter where they live. Water resistance was once even proposed as a driving force behind the evolution of feathers. That now seems unlikely, since no one believes that the first quills and down predicted by Prum's theory were waterproof. But with their intricate touch points and air pockets, vaned feathers seem surely to have been fine-tuned by this function, adapted not only for flight and display but to keep birds warm and dry in any weather.

The only noteworthy exception to the watertight rule occurs in some of the driest places on earth, where desert sandgrouse live with a whole different set of worries about water. When British naturalist Edmund Meade-Waldo first described sandgrouse breeding behavior in 1896, no one believed him. Found in arid regions from the Kalahari north to Spain and as far east as Mongolia, the various species of sandgrouse all nest on the ground, in simple scrapes or even in camel prints, often as far as thirty miles from the nearest water source. They eke out a living eating dry desert seeds and must drink regularly to survive, so adults fly round-trip to water several times every day. Meade-Waldo reported that during the breeding season, male sandgrouse would linger at the pools, wading in and methodically soaking their chests before returning to the nest. Upon his arrival, the thirsty chicks rushed out to eagerly drink at Papa's breast, sucking water straight from his feathers.

Though Meade-Waldo wrote several papers on sandgrouse and went so far as to raise the birds himself and observe this behavior in captivity, the scientific community dismissed his story as fantasy. Not only did it sound ludicrous, but it flew in the face of well-established scientific wisdom. Everyone knew that feathers repelled water; they didn't absorb it. And even if the plumes got soaked, how could they possibly remain so when flying at high speeds for thirty miles through hot desert air? It would take sixty years, repeated field observations, and an electron microscope to prove Meade-Waldo right.

Again, the answer lay in structure, a peculiar sandgrouse quirk that makes the barbules of male breast feathers (and to a lesser degree on females) grow not in a tight lattice but in loose, springlike coils. Under magnification they look something like plastic pot scrubbers, and each tiny spiral can soak up a surprising amount of liquid. Ounce for ounce, sandgrouse feathers hold two to four times as much water as the average dish sponge. Even after a long flight through the desert, a male sandgrouse in good plumage can provide each of his offspring with several beakfuls of cool, refreshing water.

I would have welcomed a blast of warm desert air by the time my murre and I reached the beach, but the breeze coming off the water had the damp, chill bite of a Pacific Northwest winter. Plunging into the wind and waves was just what the bird needed, but part of me still felt like taking the poor thing home to warm up beside the woodstove. The night was pitch-black, cloudy with no moon, and I felt my way slowly over the driftwood before finally reaching a clear sweep of gravel and sand. When I knelt to set the bird free, we released our grips on one another at the exact same instant, as if this were a dance we had rehearsed many times. In the darkness, I rubbed my thumb and listened as he scrabbled down to the safety and comfort of a frigid sea and was gone.

While the structural physics of touch points and surface tension remain debatable, the *metaphysics* of feathers and water are well known. Contemplative thought goes hand in hand with plumes and flowing water, embodied in a cultural tradition that may stretch back to the first person who ever pondered trout in a stream. Fly-fishing as sport and compulsion dates at least to the second century AD, when Greek historian Ælian famously described the angling habits of locals on a Macedonian river: "They fasten red (crimson red) wool around a hook, and fix onto the wool two feathers which grow under a cock's wattles, and which in colour are like wax. Their rod is six feet long, and their line is the same length. Then they throw their snare, and the fish, attracted and maddened by the colour, comes straight at it, thinking from the pretty sight to gain a dainty mouthful; when, however, it opens its jaws, it is caught by the hook, and enjoys a bitter repast, a captive."

The pastime was well established by the seventeenth century, when English writer and man of leisure Izaak Walton compiled his lengthy treatise and ode, *The Compleat Angler*. In it he advised every self-respecting fisherman to keep with him at all times "the feathers of a drake's head" and "other coloured feathers, both of little birds and of speckled fowl." Whether prehistoric cultures practiced the art of fly-tying is unknown, but surely people have been tricking fish with feathers since the dawn of Western civilization. As a feather-obsessed descendant of strongly piscivorous Norwegians, I knew I had to try my hand at it.

"You know, I'll need a permission slip from Eliza before I teach you how to fly-fish," John said when I called him up. Then he laughed, and I could hear him take a long drag on his cigarette. "There's a kind of obsessiveness that creeps in . . . "

As a veteran fishing guide, John Sullivan had probably risked a lot of spousal disapproval over the course of his career, aiding and abetting people's fixations on fish and feathered flies. He led trips for everything from cutthroat to small-mouthed bass, but specialized in the pursuit of steelhead, a sea-run trout so famously elusive that anglers call it "the fish of a thousand casts." People tell of long days, a season, or even years spent trying to hook and land a single specimen. But for the true aficionado, feeling the tug and pull of a steelhead take a hand-tied fly, and watching it race across the surface of a shallow western river, is an experience well worth waiting for.

John and his family live on the wooded rim of a river canyon in eastern Oregon, not far as the crow flies from my in-laws. It's sparsely populated country where anyone within driving distance qualifies as a neighbor, and he'd known my wife ever since she was a little girl. Though he'd been a professional guide for more than twenty years and people came from around the world to float rivers with him, he sounded a bit wary about taking on a client with family right down the road.

I assured him that Eliza wouldn't mind. After all, she had already let me go to Las Vegas and interview showgirls for this book—what was a fly-fishing lesson compared to that? John agreed in the end, but only after determining that I wasn't a golfer. Every person should be allowed one compulsive hobby, he reasoned, and if I didn't play golf, then there just might be room in my life for fly-fishing. (I decided not to tell him that my feather research could easily lead me to a golf course, too. Before the advent of rubber and modern synthetics, the best golf balls in the world featured a goose-feather core hand-stuffed into cowhide spheres. Packed wet, they dried into tight, hard little slugs called "featheries" that could travel more than two hundred yards, twice the distance of their wooden predecessors and still a respectable drive on most courses.)

I arrived at the Sullivan residence on an unusually chilly spring afternoon, with rainsqualls roiling up the canyon in broad, dark waves. John's boat sat in the yard on a trailer, tightly tarped against the weather. Trout season was open, he explained, but heavy rains had "blown" all the rivers. Instead of drifting downstream somewhere learning to cast, we retreated inside to concentrate on fishing's most fundamentally feathery aspect: the craft of tying flies.

"You have no idea what you're getting yourself in to," he chuckled and started spreading gear across the dining room table: two steel vices with narrow, toothed clamps and rotary arms, boxes of hooks, bobbins of thread, three pairs of Dr. Slick scissors (curved, straight, serrated), a dubbing spike, a whip finish tool, chenille, beads, tinsel, and various shades of ribbing. And then there were feathers. It was only a small portion of his supply, enough to make the few training flies he wanted to show me, but the table held dozens of loose plumes, flight feathers, and bags of colorful fluff. There was the entire brindled hackle of a rooster, as well as plumage from at least a half-dozen species, from goose and turkey to pheasant and teal. And every style bore its own fly-tying nickname: grizzly, badger, marabou, parachute, biot, cape, or saddle. A hen feather dyed an eye-popping Day-Glo yellow was "Hanford chicken," named for the famous nuclear weapons site alongside Washington's Columbia River. Certain high-quality feathers bore the name of a famous breeder, like "Hoffman grizzly," "Herbert hackle," "Metz saddle," or "Conranch stem." The specialty feather business can be lucrative, with the finest rooster skins selling for hundreds of dollars each. They're bred for specific traits—a pliable rachis and barbs that splay out evenly when wrapped. The resulting birds look as pampered as show dogs, with long, ornate plumage that gains attention far outside the fly-tying community. At Yale's Peabody Museum, Rick Prum had shown me a stuffed and mounted "Herbert Miner Cream Badger," a cock bred for fly-tying, proudly displayed among the most beautiful wild birds in the world.

The lesson passed quickly as John walked me through everything on the table, explaining each piece of equipment and showing me how the feathers could be tied and twisted around a hook shaft to mimic the wings, legs, or bodies of particular insects. A wiry, fit man of middling years, he had a salt-and-pepper beard and the permanently tanned face of a lifelong outdoorsman. He got interested in tying flies "the same way most people do—when I saw guys catching fish with them!" Though John insisted that he wasn't an expert, he belied that claim all afternoon with his wealth of detailed knowledge about the history, methods, and minutiae of feathers and fishing. He spoke with a combination of professional interest and an enthusiast's zeal, but stopped well short of the monomania he had warned me about. John has the natural curiosity of a philosopher, and our conversation veered regularly into points of geology, soil chemistry, meteorology, botany—topics likely to cross one's mind on slow floats through arid western canyons and the thousand casts between fish.

"Now you're going to tie a Silver Hilton," he informed me, and quickly demonstrated the steps, twirling thread, black chenille, silver tinsel, rooster hackle, and bits of turkey feather onto a shiny black hook braced in the vice. My attempt didn't go quite so smoothly, but fly-fishing had taught John great patience and he bore with my clumsy winding, unwinding, and rewinding until finally giving the fly his seal of approval: "I'd fish with that."

What lay before us was a fuzzy black body with a feathery ruff and shiny stripes, trailing two plumed wings and a tail. "With a dry fly you're trying to imitate an insect caught on the surface of the water," he explained, and pointed out how the barbs of the rooster hackle stood out in a spray of points that maximized surface area and kept the fly perched lightly on top of the water. It was touch points and surface tension all over again, but with a twist: dropping these feather barbs into just the right current would make them twitch and jiggle like the legs of a struggling bug.

Outside, the rain had let up, and rays of sunlight angled across the deck. A large hummingbird feeder dangled from the roof beam, and soon Rufous and Calliope Hummingbirds began arriving in droves, taking advantage of the break in the weather to fuel up on the sweetened water. I caught flashes of iridescent green backs and the brilliant crimson of a throat patch and couldn't help wondering how their feathers might look to a trout, staring up from a streambed at some bit of unimaginable fluff drifting by in the stream.

Before we called it quits, John gave me a tour of his fly collection. "This is really only a part of it," he said, pulling out box after box, each divided into dozens of tiny compartments. The bins held a dizzying array of bright, hooked baubles, as if the hummingbirds had come inside and arrayed themselves neatly in plastic trays. The obsessiveness John warned me about began to make more sense as he rattled off name after name and showed me the subtleties within each pattern. Here were muddlers, nymphs, streamers, skunks, poppers, rubber legs, duns, Hiltons, agitators, wogs, and articulated leeches, just to name a few. They ranged in size from wisps smaller than a fingernail to giant hot-pink fuzz balls as big as a boiled egg. Each design could have any number of color and texture variations—brown hackle or black, teal feathers or Mallard, a chenille body or one of rabbit hair wrapped around thread and teased with a needle. Feathers featured prominently in almost every fly—from delicate wings and tails to body fluff or long streamers dangling back from the hook. The attention to detail was phenomenal—the "Copper John" included tiny silver balls meant to mimic the air bubbles some insects carry with them underwater. "Honestly, some of these I don't even fish with," he admitted. "I just tied them because they look cool."

Perusing John's handiwork gave me a real sense of the art and craftsmanship involved in tying flies. He downplayed his own skills, however. "This is nothing. If you really want to see feathers, you need to check out the old Atlantic salmon patterns!" He directed me then to several lavishly illustrated books devoted to the famed English fly-tiers of the late nineteenth century. The period marked one of sportfishing's greatest heydays, a time when angling for Atlantic salmon was in vogue with Victorian gentlemen who strove to outdo one another with the extravagance of their lures. And just as the broad reach of empire made a world of feathers available to adorn hats, so too did fly-tiers have access to a limitless variety of colors and textures. Their creations

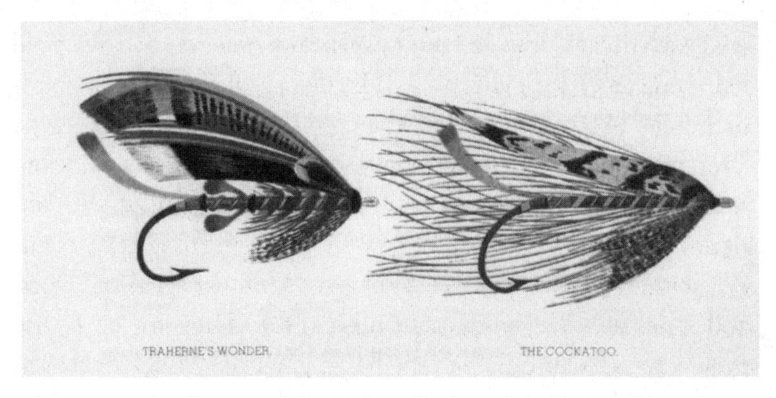

These two nineteenth-century Atlantic salmon flies featured the plumage of more than ten species from four continents, including Mute Swan, Scarlet Macaw, Ostrich, Helmeted Guineafowl, and Red-tailed Black Cockatoo.

looked less like mock insects than some exotic aviary, where the feathers of parrots, peacocks, jungle fowl, kingfishers, toucans, ostrich, and tanagers grew together in wild, hooked sprays. For some patterns, individual barbs from different species were painstakingly teased apart and then melded together to form new multi-hued feather vanes. Like some of John's favorites, many of these flies were never fished, and vintage collections now command surprising sums at auction. Even modern reproductions can fetch as much as two thousand dollars *per fly*.

For John, though, part of the fun in tying feathered flies lies not in selling them but in giving them away. He sent me home with a double handful of bright patterns designed for Pacific salmon—the cohos, Chinook, pinks, and sockeye that pass by our island each summer en route to their spawning rivers. His gift included tips on casting and how to get the right action from each fly, and then he gave me some parting words of wisdom. Even if I did everything right, he warned, I might not get a bite. "Having the perfect fly is a lot less important than actually getting it in front of a fish."

A few months later I tried to do just that. With a secondhand fly rod and a borrowed reel, I headed down to the beach at the south end of the island, not far from where I'd released my Common Murre. It was an early-autumn afternoon, warm and bright with sun. The water stretched out across the straits like smooth glass, etched here and there with tide lines and swirls of current. A hundred yards out, Surf Scoters and a Western Grebe dove and surfaced again and again, chasing minnows through the shallows. Eliza and Noah had joined me, and I could see them making their way along the beach, Noah holding her hand and tottering forward eagerly. I assembled the rod, threaded line through the guides, and proudly tied on my very own Silver Hilton. It whistled through the air by my ear as I cast and watched it settle, a speck of fluff drifting on top of the clear, cold water.

If this were a work of fiction, now is when the salmon would strike, taking the fly and surging across the surface in a series of silvery leaps. Of course, that didn't happen. My chances of actually catching a fish probably suffered from the fact that I don't know how to cast. Try as I might, I never got the damn fly to go farther than about ten feet in front of me. And after a few attempts I realized that my fly-tying skills weren't much better. The Silver Hilton had come almost completely unraveled, releasing its feathers and hackle until only a sad wisp of tinsel remained. Clearly, I had a lot more to learn about fly-fishing, but John was right about one thing—I could see how it might get addictive. I put the rod aside for a moment and just stood there, listening to the murmur of waves and the throaty squabbling of the birds. Down the beach Eliza and Noah had found a place to sit in the sun, and I watched them while he busily picked up stone after stone and tried to sneak them into his mouth. If fly-fishing embodies the metaphysics of feathers and water, then the real lesson lies in the reality of where it takes you. And if it could bring me back to the beach for another afternoon like that one, then I'd do it again in a heartbeat.

The gentlemanly art of fly-casting.

CHAPTER FOURTEEN

The Mighty *Penna*

Oh! nature's noblest gift—my gray-goose quill!
Slave of my thoughts, obedient to my will,
Torn from thy parent-bird to form a pen,
That mighty instrument of little men!

—Lord Byron,
from *English Bards and Scotch Reviewers* (1809)

The quantity of feathers in my personal collection took a sudden upturn recently thanks to our neighborhood fox. Though Eliza and I had fortified the orchard with six-foot fencing, subterranean chicken wire, and an electric shocker (and vowed never to calculate the per-egg cost of our chicken venture), the fox found an opening one night and raided the coop. He made short work of Fatty, White One, and Lurky, and while cagey old Trouser evaded him once, she had no one to hide behind when he returned a few weeks later. We were very sorry to lose them, but there was a bright side: the fox did save us the trouble of doing the job ourselves.

The American Poultry Association's *Standard of Perfection* lists both Wyandottes and Rhode Island Reds as "dual-purpose" birds, suitable for meat as well as egg production. The idea is to eat the eggs when the birds are young and vigorous, and then move the hens along to roasting pans and stew pots when their laying starts to drop off. Predictably, we'd grown quite fond of our flock and were rather dreading the second chapter of that story. In fact, the orchard stood a fairly good chance of becoming a retirement home for hens past their prime. It wasn't long ago, however, that people regularly raised and slaughtered backyard flocks of a bird that could safely be called "tri-purpose" or even "quadri-purpose."

In *The Adventure of the Blue Carbuncle*, Sherlock Holmes famously wagered with a shopkeeper that he could distinguish between the taste of a town-bred goose from one raised in the countryside. He lost the bet but did solve the mystery of how a stolen diamond ended up lodged in the crop of the bird on his sideboard. The fact that anyone would make such a wager, however, shows how common roasted goose used to be in the Western diet. Though still popular in Asia, goose meat has otherwise declined alongside the fortunes of one of its key by-products, the quill pen. Though not typically filled with jewels, a domestic goose did offer the farmer four distinct income streams. Anyone with a backyard flock could expect their geese to produce not only salable eggs, meat, and down but a handful of long, gracefully curved outer flight feathers—for more than a thousand years, the finest writing implements in the world.

Quill pens were in common use by the time that Saint Isidore, the archbishop of Seville, compiled his twenty-volume encyclopedia in the early seventh century. He noted the tools of the scribe as the reed-pen, a standard writing tool for more than two millennia, and the quill. "The reed-pen is from a tree; the quill is from a bird. The tip of a quill is split into two . . . in order that by the two tips

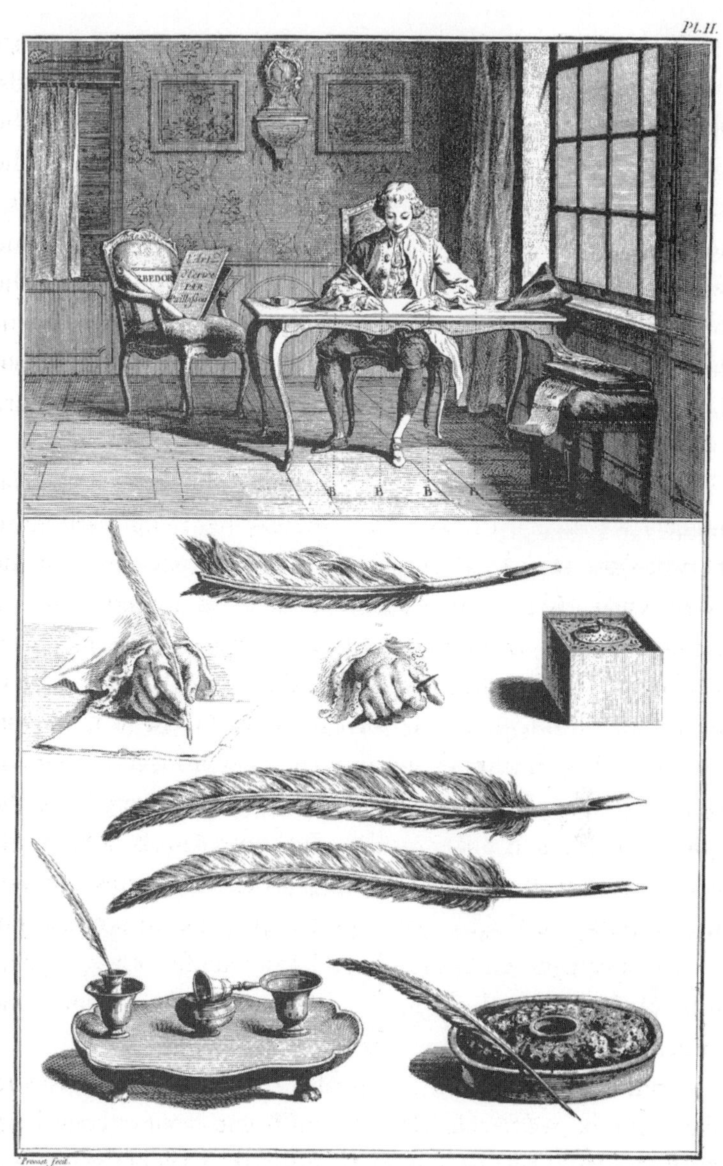

Denis Diderot devoted pages to the art of writing with quills in his eighteenth-century treatise on everything, *L'Encyclopédie*.

may be signified the Old and New Testament, from which is pressed out the sacrament of the Word poured forth in the blood of the Passion." Isidore's metaphor may sound a bit overcooked, but it tells us quite a lot. His account is considered the first definite reference to the use of a quill, and already it had supplanted its reed-pen predecessor in the weightiest task of the day, copying and illuminating scripture.

The word *pen* itself derives from the Latin *penna*, for "feather," and the penknife can similarly be thought of as a feather knife. It developed as a vital writing accessory, always kept near to hand for the constant sharpening and shaping required to maintain a steady, even line. Quills could be cut to make lines in a variety of widths, and they offered greater durability than reeds, but their real advantage lay in the curved hollow of the tube, a natural ink reservoir that dripped less and required far fewer refills while writing. (Although not made from a tree, as Isidore claimed, reed-pens were a more rigid and unforgiving tool, crafted from the cured and dried stems of certain wetland plants.) The large primaries from goose wings made up the bulk of the quill trade, though swan, turkey, and even eagle plumes were used on occasion. Smaller birds came into play for detail work, and crow feathers became so popular for technical drawing that their name lingers on: graphic artists still refer to any narrow-gauge dip pen as a "crowquill."

Production of quill pens peaked in the early nineteenth century, when education, literacy, and letter writing were on the rise but before steel nibs became widely available. Though backyard flocks were common throughout Europe, and even though some regions of Poland and Russia raised geese en masse expressly for the quill trade, supply could hardly keep up with demand. One stationer in Shoe-Lane, London, reported annual sales of six million cut and dressed quills during the 1830s, and exports from St. Petersburg alone reached as high as twenty-seven million in a year. Considering that each bird yielded only five usable feathers per wing (the largest primaries), it's no wonder that people developed a familiarity with the taste of goose flesh.

The process of making or "dressing" a good quill pen involved cleaning and hardening the tip and cutting off at least the lower portion of the vane. (The image of writing with a large, flowing plume originated in Hollywood—it makes good cinema, but in practice the barbs just got in the way and most quills were stripped clean.) Some makers boiled their feathers; others dipped them in hot sand or mild acid—the various techniques became valued trade secrets. Quill dressers in Holland perfected a technique using hot ash or sand, which came to be known as dutching. The point of a well-dutched quill took on a pearly or yellowish color "like fine, thin horn," as a reporter for London's *Saturday Magazine* described it in 1838. That same article noted how people often revered the pens that composed a famous work. Some were glazed and framed for display, while in one case a quill "was put into a golden casket by the over-zealous fan of a celebrated writer." Odes to their goose quills were not an infrequent topic for poets of the day.

By midcentury, however, mass-produced steel nibs began to flood the marketplace. Although early models were crude and wrote poorly, they sold for as little as thirty-six to the penny, one of the first truly disposable products of the industrial age. The reign of the quill began to ebb, and not everyone was sad to see it go. In Charles Dickens's weekly magazine, *Household Words*, an 1850 article gave a less than fond remembrance. It described the heretofore common scene of a schoolboy timidly addressing his writing master with, "Please sir, mend my pen."

> A slight frown, subsides as he sees that the quill is very bad—too soft or too hard—used to the stump. He dashes it away, and snatching a feather from a bundle—a poor thin feather, such as green geese drop on a common—shapes it into a pen. This mending and making process occupies all his leisure—occupies, indeed, many of the minutes that ought to be devoted to instruction. He has a perpetual battle to wage with his bad quills. They are the meanest produce of the plucked goose.

Just as the reed-pen persisted in Isidore's day, feathers remained in use alongside their successors for decades. But steel nibs, fountain pens, and ultimately the ubiquitous ballpoint prevailed, and now the quill finds its role in letters reduced to mere ceremony. The United States Supreme Court, for example, maintains an old tradition of placing twenty fresh goose quills on the counsel tables every morning. No one writes with them anymore, but visiting attorneys are invited to take one home as a souvenir.

To some artists and calligraphers, however, the rigid scratch of metal will never match the organic feel of a quill-drawn line, and a few stalwarts have kept the old techniques alive. Still, few major works had been attempted in centuries until a Benedictine abbey in Minnesota ordered a commission from calligrapher Donald Jackson in the late 1990s. It would take Jackson and a dozen of the world's best scribes and illustrators thirteen years to fulfill that order: the world's first quill-drawn, fully illuminated Bible since Guttenberg invented the printing press.

"It's like Mount Everest," Jackson said simply. "For a calligrapher, this is the ultimate challenge." He spoke with me by phone from the Scriptorium, his workshop in rural Wales. A lifelong artist and scribe, Jackson had reached the top of his profession decades before the monks of St. John's Abbey came calling. He mastered the use of a quill pen drawing up official proclamations for Queen Elizabeth II and the House of Lords. But he had always dreamed of writing and illustrating a Bible by hand. "I had it in mind as the bull's-eye," he told me. "What would it feel like to write those words with a beautifully crafted pen, the tool that all of our forebearers wrote with back into the ages?"

Not everyone working on the St. John's Bible felt the same way. "At first the scribes were upset when I required them to use quills," he recalled. "They felt they weren't getting a perfect line. We live in an age of Formica—people expect things to be flawless, without a spot. But there are other forms of perfection." He explained how the rhythm of writing was different with a quill—more sensual, and more personal. "If I dropped a feather into your hand, you would feel its touch but no perceptible weight. When you fashion a pen from a weightless material, it becomes part of you."

Eventually, every scribe on the project learned to mix their own inks and cure and shape their own pens. And the illustrators mastered old techniques for blending pigments and for fixing gold leaf onto each illumination with only the heat of their breath. The results are stunning. Each page measures two feet by three feet, a tawny sheet of calfskin vellum covered with flowing script and vivid images. Whatever blemishes the scribes had worried about are lost in the greater perfection of Jackson's vision.

When he and I spoke, Jackson was working diligently on the Bible's final chapter. It was due for delivery about the same time this book was scheduled for publication. After that, the Scriptorium team would disperse, each artist taking home a hard-earned knowledge of an ancient craft. He said he hoped they would pass it on. "Look, I'm old," he added. "I'll keep plodding along, but I'm tired!" Then he pointed out another advantage of the quill,

"Sower and the Seed," an illumination from the St. John's Bible.

one that lets a calligrapher like him keep "plodding" well into his seventies. No hand pain. Quill pens are so light, they don't need to be held tightly. "A quill invites you to caress it, not to grip it. You should be able to walk past someone writing with a quill pen and pluck it right from their fingers!"

Talking with Jackson inspired me to put my research on quills to the test. I wanted to cut my own pen, dip it in ink, and experience that organic relationship with words, the lightness and sheer sensitivity that he'd described so well. After all, wouldn't it be fitting to actually write the quill-pen chapter with a quill pen? I begged a sackful of flight feathers from a neighbor who keeps geese, scooped up some sand from the beach, sharpened my penknife, and set to work. The sand heated quickly atop the woodstove in the Raccoon Shack, and soon I was dutching quills, watching the shafts change from translucent to pearly yellow, just like they were supposed to. I studied old illustrations of a properly shaped pen and did my best to whittle out the scooped sides and split the chiseled tip. It was like carving a thick, brittle straw, with shavings flying off in springy strips. The process took most of an afternoon, but after a few false starts I finally had three pens that looked like the real thing. (This may sound like a lot of effort just to write a few lines—particularly for an author close to his deadline—but at least I wasn't trying to make reed-pens, where the curing process involves a six-month submersion in fermented cow dung.)

I put a clean sheet of paper on my desk, dipped my best quill in ink, and started to write. The tip immediately dropped a big black blob in the middle of the page and then skittered wildly away, sending a spray of dark droplets across the screen of my nearby laptop. (Take that, technology!) After several more tries and a bit of additional whittling, I started eking out a few short, smooth lines. The pen made a pleasant scratching sound as it wrote—like a fingernail quietly tapping the page in thought—and the lines had an appealing flow, like brushstrokes. It soon became clear, however, that I would not be writing any chapters with a quill, at least not in time for this book. Short horizontal lines I could manage, but the angles and curves of actual letters resulted in

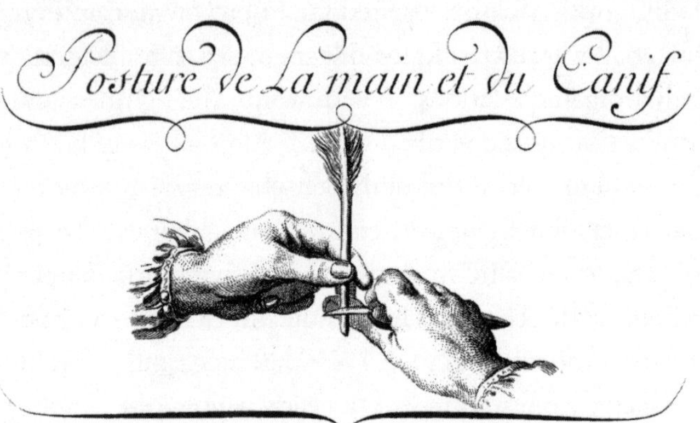

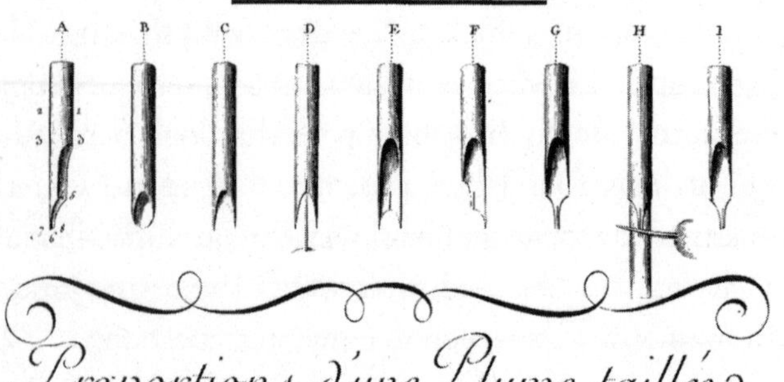

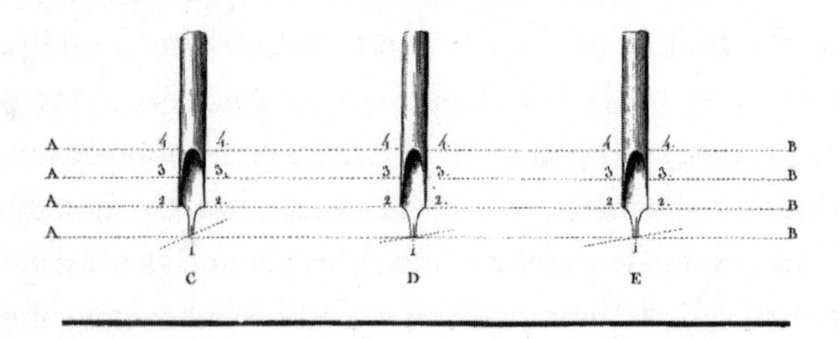

Paillasson Scrip. Aubin Sculp.

Instructions on cutting a proper quill, from Diderot's *L'Encyclopédie*.

splutters, splatters, and emptiness. I quickly revised my quill-writing goals downward: I would learn to sign my name. It's still a work in progress.

While the feathers found decorating hats or filling sleeping bags have obvious counterparts in nature, there is no evidence that birds (or theropod dinosaurs) ever copied out scripture. Nor did they dash off letters, jot down thoughts and poems, or otherwise write a word. The story of quill pens reveals that the usefulness of feathers is not at all limited by the purposes for which they evolved.

In music, the success of a song often relies less on its originator than it does on the interpretations that follow. Bart Howard's composition "In Other Words" was well received in 1954, when a singer named Felicia Sanders first performed it at the Blue Angel club in Manhattan. Ms. Sanders worked the tune into her regular act, and even put it on a record a few years later. But it's safe to say that the song, better known by its first line, "Fly Me to the Moon," is remembered more for versions recorded later by Frank Sinatra, Peggy Lee, Count Basie, Nat King Cole, Tony Bennett, Diana Krall, and—at latest count—more than twenty-two hundred other artists. Originally a jazz waltz, its various incarnations now touch on virtually every genre: swing, pop, bossa nova, Dixieland, klezmer, string quartet, and spoken word. It has been sampled by rap artists, performed on a Theremin, sung by puppets, and even recorded by a steel drum band. Mr. Howard lived off the royalties for the rest of his life.

Similarly, one measure for the merits of feathers lies in their myriad alternate uses, the countless ways they've been adapted for purposes far removed from their original functions. Grebes, for example, eat large quantities of soft body feathers and also feed them to their newly hatched chicks. This habit helps protect the birds' digestive tracts from the sharp, indigestible remains of the fish they eat—the bones collect inside a neat ball of plumes that can be safely spat up at will. A Great Crested Grebe, for example, maintains an average of eighty-seven feathers in its stomach at any given time. Most come directly from its own breast and belly, but the bird will also gobble up any suitable duck or goose plumes that happen to drift by. In several species of swallow, what apparently began as competition over nest lining has transformed into an elaborate ritual that many ornithologists interpret as play. When a suitable feather is found, pairs or groups of birds pursue one another madly, performing dramatic swoops and dives as they drop and catch the plume repeatedly in midair. Lone birds will also perform these feather chases, flying with such exuberance that even the most hard-nosed scientist can't help thinking they do it for fun.

While birds may employ feathers as flying toys or stomach remedies, the vast majority of odd uses are, like the quill pen, purely human inventions. In 1911 the magazine *Hunter, Trader, and Trapper* published an article encouraging its readers to get involved in the lucrative American feather industry. It noted the usual markets for down and millinery plumes, but also listed "pens, featherbone powder puffs, trimming, boas, 'furs,' fans, parures, military and lodge plumes, fire screens, artificial flies for anglers, brushes, tooth picks, dusters, camel's hair brushes and even, though rarely, parasols." Of these, the feather toothpick had grown from a humble way to recycle worn-out quill pens into a surprisingly widespread and profitable enterprise. The habit of picking one's teeth with a feather dates to at least Roman times, but in the nineteenth century quill toothpicks emerged as one of the first mass-produced tools for dental hygiene. Several could be made from a single feather shaft, with one end sharpened and the other shaped like a scoop. Sanitized, individually wrapped, and occasionally flavored, quill toothpicks were available in restaurants, hotels, and drugstores and sold on street corners in cities around the world. They remained a viable competitor to wooden picks until well into the twentieth century.

While the market for quill toothpicks and featherbone powder puffs may have faded away, engineers and entrepreneurs continue to find surprising new uses for the plumage of birds. Promising experiments appear regularly in a range of scientific journals, but the best place to look is in the records of the U.S. and European patent offices. They both maintain extensive online databases where any simple search reveals scores of sometimes bizarre ideas for the next great feathery product. If one group of inventors has their way, we may one day fuel our cars with feathers. A recent study found poultry waste to be an excellent source for the production of biodiesel, with an estimated global market of several billion dollars a year. Patents for feather fabric range from a 1902 version featuring down mixed with wool and cotton to a yarn from turkey barbs to a polyester-like fiber reconstituted from a goopy feather slurry. There are patented ostrich-plume air purifiers, feather-stuffed orthotics, biodegradable feather plastics, feathers for erosion control, and feather-based electronic circuit boards. Bacteria that produce antimalarial biocides grow well on processed feather keratin, and traditional feather dusters are now available in touch-activated, vacuum, and aromatherapy models. Feather barbs can be processed into notebook paper, insulation, and upholstery padding, as well as an absorbent fiber that shows great promise for use in biodegradable baby diapers.

Writing the preceding paragraph, I felt a bit like a huckster plugging some miracle product on late-night television—"But wait, there's more!" Although it's true that the great variety of feather structures in nature translates to a huge potential for artificial uses, it's important to remember that most of these schemes will never make it past the laboratory, drafting table, and patent office. And it's also important to keep in mind that feathers aren't entirely a wonder product in the natural world, either. In fact, for some birds, in certain situations, they're a huge disadvantage.

CHAPTER FIFTEEN

The Featherless Head

The Vulture eats between his meals,
And that's the reason why
He very, very, rarely feels
As well as you and I.

His eye is dull, his head is bald,
His neck is growing thinner.
Oh! what a lesson for us all
To only eat at dinner!!

—Hilaire Belloc,
More Beasts for Worse Children (1897)

"Peck, jab, tear." I paused to wipe sweat from my eyes, then squinted back through the scope. "Jab, hop, hop, tear. Swallow!"

Beside me, Diana scribbled down the data, while the third member of our team, another Diana, scanned ahead with binoculars. "Two Nubians," she called out, "three Egyptians . . . twenty-four White-Backs . . . fourteen Ruppells and seven—no, eight—Marabou Storks."

We crouched in the small shade beside our Land Rover, alone on a dusty flat plain dotted with ant-ball acacias and fever trees. Fifty yards ahead of us, the ground seethed with vultures. They crowded together, shoulder-tight, a mass of hunched brown backs and bobbing heads jostling for position around the carcass. We could hear their hissing and the snap of beaks.

Through the spotting scope, I watched my bird rear up and leap twice through the melee ("Hop, hop!"), lunge at a neighbor ("Jab!"), rip something red from the pile ("Tear!"), and lift its head high in a jerking, curve-necked gulp ("Swallow!").

"Time," someone called, and I sat back, grateful for the break. We'd been at it for hours under a scorching Kenyan sun, picking random birds from the flock and shouting out their every behavior. We were studying the feeding hierarchy of four different vulture species, as well as the hulking Marabou Stork. The idea was to relate units of effort (jabs, tears, hops) to units of reward (swallows of meat). How many jabs did it take to get a swallow? Did smaller species expend more energy to feed? How did these relationships change over time, as a carcass diminished from a whole antelope down to entrails, skin, sinew, and bone? Such are the questions that keep vulture researchers up at night.

In the end, the project didn't amount to much. It was hard to maintain a steady supply of dead animals, and even harder to pick out the subtleties of a vulture's jabs and swallows in the midst of a throng of shifting, flapping bodies. But the experience did leave me with an intimate understanding of one small evolutionary puzzle: why a vulture's head is featherless.

I noticed the two Dianas keeping their distance, staying as far from me as possible without leaving the shade of the vehicle. I couldn't blame them. My hair, face, and forearms gave off a sickening boneyard stench as rank as anything drifting our way from

White-backed Vultures in Kenya.

the vultures' direction. I smelled like a rotten zebra, and for good reason.

Earlier that day we'd stopped by the slaughterhouse where we got most of our carcasses. They specialized in game animals and often worked at night, processing the culls from a nearby ranch. (The meat was sold to a popular Nairobi restaurant called the Carnivore, where tourists rounded out their safari experience with a taste of the veld: giraffe burgers, barbecued gnu, that sort of thing.) The butchers set things aside for us and usually had a small trailer ready in the morning, neatly packed with vulture snacks. That morning, however, we'd found something different—an unsorted heap of organs, intestines, hooves, and zebra bits piled head-high in the side yard.

"*Karibu sana,*" the old watchman told us. The shop was closed for a holiday, he explained, but we were welcome to help ourselves.

Unfortunately, we lacked a shovel, pitchfork, gloves, or any other practical means of heaving the carrion into the trailer. Even worse, the material was a day or two old and had quickly gone to rot in the tropical heat. The yard's foul air hummed with thousands of fat green flies that bumped against us in their haste to reach the meat. I looked at the Dianas, and they looked back at me, all three of us wide-eyed with horror. We were students on a study-abroad course. We knew that field biology could be rough work, but *this* had definitely not been in the brochure.

They say the chivalrous man is out of fashion, but I find he still persists in certain situations: opening car doors and holding umbrellas on a romantic evening, helping an elderly woman with her shopping bags, or plunging one's bare hands into a pile of rotten animal intestines. Waving the Dianas back with a false show of confidence, I leaned in to the pile, grabbed hold of something, and yanked.

The cecum of a zebra is a big, purplish pouch at the head of the large intestine. This one was bloated, stretched tight like a thick, wet balloon. When it burst, rotten stomach gas exploded in my face, blowing my hair back and coating me with a spray of old blood, strands of ropy goo, and flecks of half-digested bush grass. The smell was indescribable.

I realized later that in that moment, head down, tearing into a pile of carrion, I was very much like a vulture myself. But with my long ponytail, T-shirt, wristwatch, and hairy forearms, I wasn't very well adapted to the lifestyle. The gore stuck to me everywhere, matting my hair and drying to a fetid red patina as the day wore on. I got the trailer loaded, and we got our data, but it would take three bucket showers that evening before my colleagues let me into the mess tent.

If blood and guts stick so well to hair, what do they do to feathers? To answer this question, I recently tried a little experiment. Here at home I don't have ready access to a slaughterhouse or a flock of African vultures, but the demise of our laying hens had provided me with a large supply of feathers. With a couple of dead bullfrogs and a hairdryer, I figured I could come up with something.

The bullfrogs came fresh from our pond, where I'd been stalking them for months with a pump-action BB gun. I don't normally hunt frogs, but these were an aggressive invasive species that preyed on our native amphibians, and even ate small birds. They had to go, but at least I now had a use for their remains. A short foray through the reeds netted me two good-size specimens. I gutted them into a metal bowl, added chicken feathers, and stirred.

For this project I'd specifically chosen contour feathers, the curved, soft type that would normally cover a bird's head. After a few quick passes of the spoon, they became gummy and matted, their once graceful barbs stuck to one another haphazardly. I plucked them from the bowl and fired up the hair dryer to simulate a hot savanna wind. Within minutes the feathers were dry and stiff with blood and viscera. I tried preening the barbs apart with a needle but it was hopeless—they were a perfect mess. In Kenya, my simple straight hair had resisted all shampoos and reeked for days; I couldn't imagine cleaning something as intricate as feathers.

Naked bird heads rarely appear in sonnets, odes, or love songs, but in the context of carrion they're a truly beautiful adaptation. Bare skin collects far less gore than complicated featherwork, and staying clean reduces a scavenger's exposure to bacteria, parasites, and disease. When every meal is a mess of guts and decay, it's easy to imagine natural selection favoring baldness; the bareheaded birds would feed with less risk of illness, increasing their survival and reproductive success. The message is clear: feathers may be an evolutionary marvel for flight or insulation, but they're worse than useless when you're neck deep inside a dead zebra.

For carrion birds, the loss of feathers is such a good idea that it has evolved at least twice, in different places, in totally different groups of species. Taxonomists call it a textbook case of *convergent evolution*, where unrelated creatures develop similar traits in response to similar environmental cues. The Kenyan birds I studied all belonged to the Old World vultures, a diverse group closely related to hawks and eagles. Our carcasses attracted Egyptian Vultures, two species of griffons, and the hulking Nubian Vulture, largest of all the Old World species. At the time it never occurred to me to bring a Nubian home, but if I had, its wrinkled red head and glossy black plumage would have fit right in with a flock of Turkey Vultures or California Condors. The American species, however, belong to a totally different family of birds more closely linked to storks. The New World and Old World vultures are not related; their likeness evolved from the practicalities of their grisly diet.

A vulture's featherless head reminds us that the products of evolution are never static. Even sophisticated traits remain subject to the constant refinements of natural selection and the vicissitudes of genetic drift. They may persist for eons, but only so long as they are useful or, at the very least, benign and fortunate. The feathers that lined an *Archaeopteryx* or *Anchiornis* wing looked modern in nearly every respect. In the millions of years since, feathers have become perfected for flight, fluff, and fancy, not to mention waterproofing and all the odd uses that people have come up with. But one important question remains unanswered: where are they headed now? Are there new or unknown feather capabilities still evolving?

We've seen that feathers can be lost when they put birds at a disadvantage, but vultures are far from the only species to shed them, alter them, or reshape them toward new purposes. Wild Turkeys, ibis, cassowaries, and spoonbills all have bare heads to attract mates, using brightly colored skin and showy wattles to augment their breeding plumage. Owl flight feathers feature unique comblike barbs and a trailing fringe that alters turbulence to dispel sound, muffling their every wing beat. For Nighthawks and Whip-poor-wills, long facial bristles act as an insect net and also serve the same sensory functions as a cat's whiskers. It's true that feathers have been feathers since the days *Archaeopteryx* soared over Devonian swamps, but their evolution has never been idle. A survey of modern birds finds myriad variations on the theme, from subtle tweaks to glaring modifications that help birds adapt in sometimes surprising ways.

Dr. Kimberly Bostwick works in a cramped laboratory at Cornell University's Museum of Vertebrates, poring over high-speed video footage of tiny tropical songbirds and clamping their feathers into a contraption called a laser Doppler vibrometer. A protégé of Rick Prum, Bostwick burst onto the scene in 2005 when her dissertation research appeared in the journal *Science* and made news copy around the world. For a graduate student, this is the equivalent of being called up to the major leagues and hitting a grand slam on your first at-bat. But Kim Bostwick is an exceptional scientist, and in the Club-winged Manakin she found an exceptional subject: a bird that plays the violin.

"It was the craziest thing I'd ever seen," Kim told me, recalling her first field observations on a trip to Ecuador in 1997. "I kept thinking, 'How did this evolve? What steps led to the weird world this bird is living in?'"

What Kim saw in that Ecuadorian rain forest was a tiny russet and black bird, flitting about on a branch and pausing periodically to thrust its wings skyward at a sharp angle above its back. Streaked with white, the dark wings made a striking visual display, but it was the sounds that drew her attention: sharp, metallic clicks followed by a brief sustained note she later described as a "ting." I've watched Kim's videos many times, and even played along with the bird on our piano. (For the record, the Club-winged Manakin performs its breeding display in the key of F-sharp.) To me the wing note sounds like the ring of crystal, or a quick bow stroke on a fine instrument—clear and steady, with a touch of vibrato.

A male Club-winged Manakin's striking visual (and sonic) feather display.

Like birds of paradise, manakins follow the lek model of courtship, with brightly adorned males staging elaborate dances on their display grounds. Females choose, and in this case they appear to like a bit of percussion with the usual song and dance. At least eleven distinct feather-snapping motions have evolved, and some species use more than one—beating wings against wings, body, or tail, or sometimes uncoiling them into the air like a whip crack. Combined with loud vocalizations, and faster-than-sight flight patterns, feather snaps make manakin displays one of nature's most complex (and frenetic) rituals. Among the family's sixty species, however, only the Club-winged Manakin has taken wing snaps a step further and added a string section to its orchestra.

"I was into biology and animal behavior, but I wasn't really a bird person," Kim explained, recalling her undergraduate days at Cornell. She traces her interest in manakins to three critical events. First came The Functional Morphology of Vertebrates, a college course she signed up for more or less on a whim. "I didn't actually know what it meant," she admits now with a laugh, "but I understood that it had something to do with animals and diversity." In fact, the class focused on comparative anatomy, following the evolution of specific physical traits through different groups of animals. "We learned how the gill arch of a shark was the same as an ear bone in mammals, or how fins and wings are related. I absolutely loved it." The course sparked a lasting interest in the evolution of structure and laid the groundwork for all of her future feather research.

Kim's next break came in accepting a job at Cornell's Museum of Vertebrates immediately after graduation, where she found herself suddenly immersed in birds. The museum collaborates closely with Cornell's renowned Laboratory of Ornithology, a leading center for bird research and conservation. "I thought I knew something about nature, but then I realized there were all these incredible species right in my backyard that I didn't know anything about!"

The birding bug bit her, and bit hard. She ended up moving to the American Southwest and basically devoting herself to birdwatching full-time for three years. When she finally got around to looking at graduate programs, one of the first professors she contacted was none other than Richard Prum. Prum had cut his ornithological teeth studying manakins in the 1980s and early 1990s, so he had a ready answer for a prospective student interested in comparative anatomy: "There's this group of birds called manakins.

They make sounds with their wings, and no one knows how they do it!" Kim's transition to a life of manakins was complete.

She started off in the lab, examining every specimen she could get her hands on. "I basically spent an entire year looking at wings." Even before she saw it in the field, the Club-winged Manakin stood out from the crowd. The feather shafts of the secondaries looked oddly swollen, and one of them had a bizarre forty-five-degree bend near the tip. Even the musculature was different from other bird wings. "There was definitely a good dissertation there if I could figure out what was going on," Kim said, and she remembers Prum's comment: "I don't know what those birds are doing, but it's got to be cool!"

Kim and Rick were hardly the first biologists to notice the Club-winged Manakin's feathers. The bird's very name referred to its peculiar clublike shafts, and Darwin himself used them to illustrate his ideas about sexual selection in 1871. He noted that only males sported the abnormalities, and though he never saw a manakin in the wild, he surmised that the odd feathers must be used for mate attraction: "In the case of the modified feathers by which the drumming, whistling, or roaring noises are produced, we know that some birds during their courtship flutter, or shake, or rattle their unmodified feathers together; and if the females were to be led to select the best performers . . . by slow degrees the feathers might be modified to almost any extent." Darwin intuited the purpose and the process, but exactly how the feather noisemakers worked had been a mystery ever since.

After her first trip to Ecuador, Kim was still stumped. "I came back to the lab, looked at the specimens, and decided that what I had seen was impossible." She examined the wings again and again, holding them up exactly the way the birds in Ecuador had, but it seemed impossible that the feathers could produce a tone. Snaps and rattles, yes, but the source of the "ting" remained a

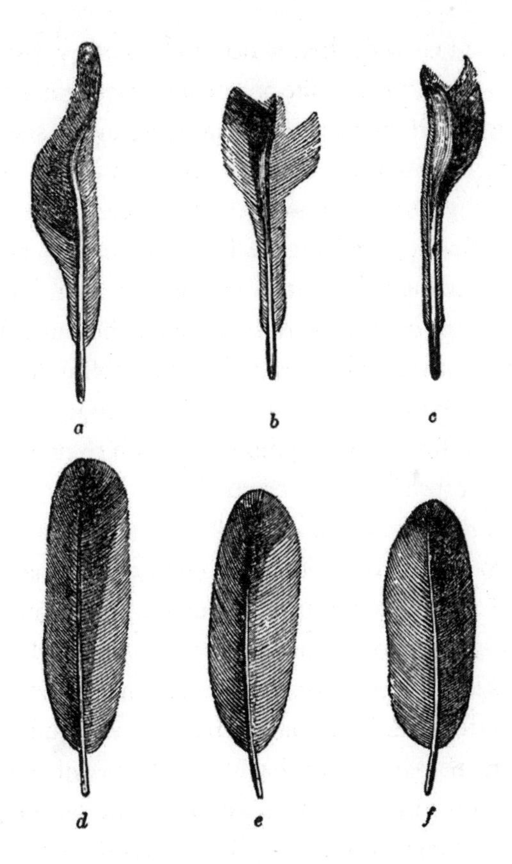

Charles Darwin correctly attributed the odd shape of the male Club-winged Manakin's wing feathers (*top row*) to sexual selection.

mystery. "There was no way the wings could be making the sounds the wings were making. We were up against the laws of physics."

Traditionally, scientists credited birds with two basic ways to make feather sounds: they could beat things together or use a rapid flight dive to set up some kind of vibration. Think of them as snaps and swoops. Snaps include the various percussive notes found in other manakins, as well as the familiar wing-clapping racket of a pigeon taking flight. Swoops occur in a range of noisy flight displays, all relying on the great speed of a dive to induce some kind of whistling or rattling. The effect ranges from the buzz-chirp of a diving Anna's Hummingbird to the eerie winnowing of Common Snipes and Nighthawks. Clearly, the Club-winged Manakin was doing something new, but like every biologist before her, Kim began to think she might never figure it out.

Then came an innovation that all her predecessors had lacked: high-speed digital video. For guerrilla filmmakers and low-budget documentarians, the advent of compact digital video equipment opened up a new world of opportunity, producing professional results at a fraction of the cost of traditional film. For scientists, it opened up a new world of potential data and unprecedented observations. Kim returned to the rain forest with a portable camera that could capture the birds' lightning-fast movements at up to one thousand frames per second, more than thirty times the rate of normal video footage.

"High-speed video was the real breakthrough," Kim told me. Suddenly, she could see exactly what the birds were doing, watch the play of every wing beat, and match each movement to its sound, frame by frame. Her videos revealed how the manakin's inverted wings "ticked" together loudly—a traditional snapping mechanism—before beginning an odd controlled shivering motion. This rapid vibration brought the wings together repeatedly, striking the enlarged clublike secondaries together in a way that forced the bent one to saw back and forth across a row of tiny ridges on the adjacent shaft. Suddenly, the whole story fell into place. Each wing was indeed acting like a tiny violin, with the bent feather tip serving as the pick or bow, the ridges as strings, and the swollen, hollow feather shafts as the resonating chamber, amplifying and sustaining the tone.

"This mechanism is totally unique," Kim explained. No other known bird makes that kind of sustained structural tone, rubbing body parts together in what scientists call *stridulation*. "The closest analogue is probably the cricket," she added, and I thought immediately of Aesop's classic fable "The Ant and the Grasshopper." In it, the carefree grasshopper saws away on his fiddle while the industrious ants busily store food for the coming winter. Like crickets, grasshoppers use a pick-and-file system to stridulate, rubbing their hind legs against their wings to create that familiar chirring backdrop of a summer afternoon. The moral of Aesop's fable breaks down in a biological setting, however, since many species of ants stridulate, too. They could have brought their own fiddles to the party and gone hungry in winter right alongside the grasshopper. In fact, similar "pick-and-file" systems have evolved in a wide range of insects, from beetles to millipedes. "If the Club-winged Manakin was a bug," Kim laughed, "it wouldn't be that interesting!"

Kim's work highlights an extreme feather adaptation that is both new to science and new in the long history of feather evolution. It occurs in only one recently derived species of bird. But there was another surprise in store: all kinds of feather shafts have the ability to resonate, even if they lack the manakin's special club-shaped chambers. What's more, every feather Kim hooks up to her laser Doppler vibrometer appears to resonate at exactly the same pitch. "It's not that manakin feathers evolved to have a resonating property. It's that they pulled out a preexisting trait and capitalized on it."

With that in mind, I asked Kim if other birds were taking advantage of the resonant quality of feathers. Even if they lacked the manakin's pick and comb, couldn't a sound produced by snapping or swooping trigger the right frequency to resonate within the feather shaft?

"I don't think anyone knows the answer to that yet," she told me, "but my guess is yes." Her own research is now heading in that direction. In a recent e-mail she said she'd found strong evidence of feather resonance in the booming wing- and breast-feather swooshes produced by male Sage Grouse. And if the potential for feather sounds and resonance is as old as feathers themselves, then perhaps Xing Xu will uncover evidence of a stridulating, snapping, or swooping theropod. "We should definitely be looking at the fossils," Kim agreed.

My last question for Kim was one I'd carried with me and puzzled over repeatedly throughout the research for this book: where are feathers headed now? Given the intricacies of their structure and development, their endless variety of colors and shapes, what else is possible? What new uses will evolve?

"The answer is . . . they're already doing it!" she answered immediately. "Feathers are already being used for things we don't appreciate. We've overlooked so much about them because they're moving too fast, or we don't understand the physics or the chemistry." Just as new video capabilities revealed the Club-winged Manakin's remarkable story, Kim believes new technologies will open up insights into feathers' insulative qualities, their waterproofing, their colors, their aerodynamics. "Feathers are already incredible in every way!" she exclaimed at one point. "We just need to be able to see it."

Like so many of the people I spoke to about their work with feathers, Kim Bostwick loves her job. She talks about science with unabashed excitement, as if each new idea is a bright package she has just unwrapped. Our first conversation was over a scratchy Skype line with a delayed connection, on a morning after Eliza and I had been up all night in shifts with a vociferously teething Noah. Yet by the end of the conversation, Kim's enthusiasm had me fully recharged and energized, ready to go out and buy a laser vibrometer or a high-speed camera and point it at the first natural miracle to flap, swoop, or flutter past the window. In a results-oriented culture like ours, it's easy to get hung up on endings, on figuring things out and finding precise solutions. But a true fascination continues building with each new piece of information—making new connections, revealing new patterns, and opening new perceptions. The exploration of natural miracles is a fundamentally open-ended and curiosity-driven enterprise. It reminds us that science is not always about the answers; it's about the questions.

CONCLUSION

A Debt of Wonder

He sees that this great roundabout—
The world, with all its motley rout,
Church, army, physic, law,
Its customs and its businesses—
Are no concern at all of his,
And says, what says he?—Caw.

Thrice happy bird! I too have seen
Much of the vanities of men,
And sick of having seen 'em,
Would cheerfully these limbs resign
For such a pair of wings as thine,
And such a head between 'em.

—William Cowper, *The Jackdaw* (1782)

A brisk wind cut the air, raising white-capped waves in neat rows that stretched away across the straits to Canada. As we hiked toward the point, bright sunlight streamed down through the evergreens, urging us ahead into its warmth. Spring had arrived early, with honeysuckle already leafing out and the first currants in bloom, their flowers dangling in the breeze like tiny pink lanterns. Noah shifted against my chest and slept on—he was already a veteran hiker, welcoming every walk in the woods as a prime napping opportunity. We were visiting Eliza's extended family, who live on an island even smaller and more remote than our own. Her aunt led the day's expedition, everyone's first chance to inspect a new trail through the forest at the island's northernmost tip. Just before we stopped for lunch, she spotted an unusual feather hung up in the branches beside the path. Plucking it down carefully, she handed it back to Eliza, who passed it on to me with a questioning look. "Any ideas?"

The feather was large, longer than my outstretched hand, with a wide, plumy base that narrowed to a small vane near the tip. The color scheme puzzled me: snowy white barbs grading to cream and buff halfway up, then tinged with a distinct reddish color. It obviously came from the breast or flanks of a large bird, but what? A goose? A raptor? Some gigantic owl? Nothing I could think of had quite the right color pattern. "I don't know what it is," I said finally, tucking it away. "But I know how to find out."

A small boat, two planes, and a subway ride later, I walked across the National Mall in Washington, D.C., heading for the great dome and Corinthian facade of the Smithsonian Institution's Museum of Natural History. Dr. Carla Dove met me "by the elephant," a self-evident gathering point below the upraised trunk of the huge pachyderm that presides over the museum foyer. Dressed casually and with a direct, unassuming manner, Carla gave no outward sign of being one of the world's preeminent feather experts. Perhaps no one alive has seen, handled, and examined a wider variety of plumes than Carla. "Come on," she said, her voice graced with the long vowels of Virginia. "I'll take you up."

We walked behind the elephant, through a security door, and into a labyrinth of halls and stairways that ended at the Feather

Identification Lab, where Carla presides over three full-time staff people, several fancy microscopes, a gene sequencer, and a collection of more than 650,000 stuffed birds. "We've got more than three-quarters of the world's species," she told me as we passed row upon row of tall cabinets, "and it's still growing."

The specimens date back to collecting trips in the early nineteenth century and include contributions from such luminaries as John James Audubon and Theodore Roosevelt. Professional feather identification, on the other hand, got its start in the fall of 1960, after the mysterious crash of a Lockheed Electra turboprop airliner.

Sixty-two people died when Eastern Airlines Flight 375 veered off course and plunged into Boston Harbor shortly after takeoff. At the time, it marked one of the deadliest plane accidents in American history and shocked a nation still enamored with the idea of commercial air travel. When investigators found bird remains—known in the trade as "snarge" or simply "bird ick"—clogging the ruined engines, the nascent Federal Aviation Administration (FAA) developed a sudden interest in the safety risks posed by bird strikes. If they could somehow identify the species involved, they could begin designing and managing airports and flight patterns to mitigate the danger.

Shortly thereafter, a messy package of snarge arrived at the Smithsonian, where it quickly found its way to taxidermist Roxie Laybourne in the Division of Birds. Using only her wits and an obscure 1916 publication on feather structure, Roxie had developed precise methods for identifying feathers from the microscopic details of their plumes. The snarge from Boston provided plenty of material, and she returned a quick verdict: European Starlings. The case was closed, the FAA was impressed, and the Feather Identification Lab was born.

"I'd never met anyone like Roxie," Carla told me soon after we sat down in her office to talk. "When I first got here, I just followed her around, learning everything that I could. And my job was in collections—I wasn't even supposed to be doing feather work!"

That was twenty years ago, and Carla has never looked back. She wrote her doctoral dissertation on the microscopic structure of shorebird feathers and joined Roxie in the ID lab full-time. "Later, when Roxie couldn't make it in to the office, we used to work cases together on her porch," Carla said, recalling the time just before her mentor passed away, at age ninety-two. Since then, the case-load at the lab has continued to grow, from three hundred identifications a year when Carla started to more than five thousand today. She and her colleagues still use the methods that Roxie pioneered, as well as new DNA fingerprinting techniques to help with fine distinctions, or when the snarge lacks any usable feather pieces. Their clients have included everyone from the FAA to the U.S. Air Force, Navy, and Army; the U.S. Fish and Wildlife Service; the National Park Service; the U.S. Bureau of Customs; the National Transportation Safety Board; and the Federal Bureau of Investigation (FBI). "It's like detective work," Carla explained. "I love this job."

And just like a detective, Carla can never be sure where a case might lead. "We recently had a deer strike at fifteen hundred feet. It was right after Christmas," she told me with a straight face. "The DNA from the snarge all checked out—it was definitely a deer." Only when investigators returned to the plane did they locate a tiny plume fragment from the unlucky vulture whose last meal led to the confusion. "We've gotten frogs that way, too, and snakes—anything a raptor might have in its claws or its stomach." Though bird (et al.) strikes remain their most common cases, the lab deals with all kinds of feather conundrums. When biologists in the Everglades catch an invasive African python, Carla can tell them what rare birds it's been feeding on. Anthropologists, museums, and Native American tribes have all sought help identifying the plumes on ancient artifacts, and the FBI once needed to match feathers in the brain of a murder victim to the pillow that had been used to silence the gun. The caseload in illegal wildlife trafficking grew so large that the Fish and Wildlife Service finally stopped sending feathers to Carla. They instead sent agents, who trained at the Smithsonian for months before returning to open their own feather forensics lab.

I asked Carla if she would walk me through the identification process from start to finish, using the strange plume I'd brought from home. "I'd be glad to," she agreed immediately, but then burst out laughing when she saw the feather. "Ooh," she said, and shook a finger at me, "you're in trouble now!"

Uh-oh, I thought. *What is this thing?* But of course that was exactly the question I'd come to answer. Carla ran the feather through her fingers in a knowing way, and then we got down to business. First of all, it was a whole feather, a semiplume, and obviously from a large bird. "And it has an afterfeather," she added. "Not all feathers do." She showed me the place in the calamus where we might find usable DNA, and then we made a microslide of two downy barbs from the feather's base. "All the useful microstructures are in the plume barbules," she explained. Through a microscope I could see what she meant. The barbules glowed under magnification like wavy glassine threads, with distinctive swellings at the nodes where their long cells came together. Some nodes were wide, some narrow, some triangular, some bore spines. We compared my plume to a variety of reference slides and quickly narrowed things down. It was not a duck or a swan. It was not a game bird. It was not an owl. Everything pointed to a particular group of large raptors.

"Now we go to the collection to confirm it," she said, and led me back to the endless rows of specimen cabinets. Opening doors

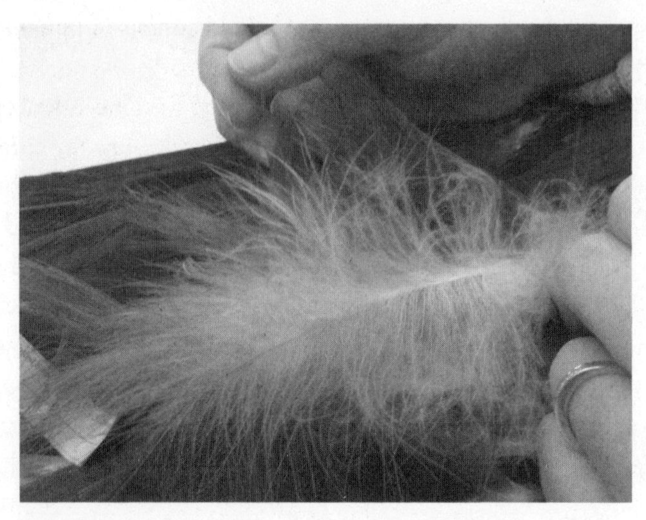

Carla Dove examines the plume I brought to the Smithsonian Feather Identification Lab.

and pulling out drawers from memory, she plopped a stiff Bald Eagle and a Red-tailed Hawk into my arms in rapid succession. Then she gathered a few others, and we laid them side by side on a central table, where banks of custom lightbulbs mimicked direct sunlight. The hawk plumes were clearly too small, the Bald Eagle's too dark. Then we found a perfect match, from the light downy base to the reddish tip. "I like this, I like this," Carla said, and held my feather up next to its twin. Apparently, I'd been walking around with a semiplume from the left flank of a juvenile Golden Eagle tucked into my luggage, in clear violation of several federal laws. After I snapped a picture and took some notes, we put the specimens away and headed back to Carla's office. She didn't give me the feather back.

As our conversation wound down, I asked Carla whether she was a bird-watcher. "Yes, I am!" she exclaimed. "I've got to get out there this weekend and see what's around." She was, in fact, an active local birder and had recently helped confirm a new species (Great White Heron) for the state of Virginia.

While researching this book, I asked that same question to dozens of people—from paleontologists and museum curators to engineers, calligraphers, and hatmakers. Almost everyone regarded it as a curious notion. Sure, the ornithologists said yes, and I did find a Hollywood fashion designer who scattered corn for the doves on his patio, but to most people, bird-watching sounded downright eccentric. In spite of a common fascination, even obsession, with feathers, we too often forget to appreciate them in their natural setting, gracing the wild creatures around us. Whether our amazement springs from the plume in a hat, the warmth of a down jacket, or the uncanny physics of a feathered wing in flight, we owe the debt of our wonder to birds.

As feather fanatics (and you now qualify, dear reader, having come this far), we needn't all be diehard birders, with cross-referenced life lists and binoculars worth more than our cars. But we must at least become bird advocates, or we risk witnessing the steady ebb and loss of the very objects of our interest and desire. As early as the 1850s, Alfred Russel Wallace noted that birds of paradise "were much more difficult to obtain than they were even twenty years ago" and blamed the trend on overhunting. What would he make of today's world, where more than one in eight bird species is considered threatened with extinction and his beloved Malay Archipelago has lost more than 70 percent of the primary rain forests he came to know so well?

Widespread, visible, and with a growing network of ardent observers, birds serve as a living barometer for many larger environmental trends. Monitoring migratory species makes it possible to track habitat loss in their winter or summer habitats half a world away, for example, while range shifts and changes in nesting season give us immediate insight into the effects of global warming.

As the footprint of human activity grows, more and more birds are becoming rare, and even common species appear to be declining. This stress on bird populations compounds the consequences of our fascination with feathers. When the plume boom of the nineteenth century drove Snowy Egrets and other species to the brink of extinction, the human population stood at 1.5 billion. If demand for wild feathers peaked again, how would birds fare now that nearly five times as many people crowd the planet? Of course, better legal protections are now in place and domestic feathers fill the bulk of the market demand, but a thriving illicit trade persists just below the surface.

Every year, millions of wild birds are hunted or captured around the world to meet both local and international appetites for bush meat, pets, feathers, fetishes, and avian curios. In Brazil alone, the illegal wildlife trade brings in an estimated one billion dollars annually, with bird and feather products making up a sizable portion of the pot. In her book *Animal Investigators*, Laurel Neme details the long pursuit of one importer in Florida who specialized in Brazilian Indian featherwork. At the time of his arrest, this single individual held thousands of artifacts and bulk feathers from rare and endangered birds, including "Harpy Eagles, Scarlet Macaws, Hyacinth Macaws, Red and Green Macaws, Blue and Yellow Macaws, Green Ibis, Guianan Red-Cotingas, Spangled Cotingas, Maguari Storks, Great Egrets, Orange-winged Parrots, Mealy Parrots, Bare-necked Fruit Crows, Laughing Falcons, Chanel-billed Toucans, Aracari Toucans, Oropendulas, Curasows . . . "—the list goes on. But one needn't track down a trafficker to glimpse the extent of the wild-bird feather trade. On any given day, a simple eBay or craigslist search will net scores of questionable offerings. I paused while writing this paragraph and quickly found feathers from dozens of unusual wild species openly for sale online, from riflebirds to trogons, barbets to turacos, and even a Greater Bird of

Paradise. Collectors and classic fly-tying enthusiasts sometimes pay thousands of dollars for hard-to-find feather specimens, and even run-of-the-mill sales perpetuate an unnecessary burden on wild populations. After all, Jodi Favazzo can custom dye chicken feathers to any color of the rainbow, and a fish that won't bite on Herbert Miner's Cream Badger rooster hackle probably isn't worth catching.

The Smithsonian's Feather Identification Lab provides an ideal end point for this exploration of fluff, flight, and fancy. Carla and her colleagues live and breathe feathers every day. Part biology, part forensics, their work occupies a unique place at the interface of the human world and the feathered world, where people and birds quite literally collide. When their efforts succeed, when the snarge and feather barbs yield a positive ID, they demonstrate how the impacts of colliding worlds can sometimes be softened. Wildlife crimes can be solved, traffickers can be caught, and airports can be redesigned so that fewer birds are killed and fewer people put at risk. It's a lesson we need to take with us to all the flash points of conflict between human and natural systems. How we choose to resolve those problems now, from habitat loss to invasive species to climate change, will determine exactly what measure of biodiversity our generations leave for those that follow. Which warblers and bee eaters will survive? Which owls, thrushes, gnatcatchers, auks, swifts, and eagles will persist in the wilds of their habitat, and which will be found only in museums?

At the end of our meeting, Carla escorted me back out to the elephant in the foyer, and we shook hands good-bye. The museum was in full swing, with families, tourists, and groups of chattering schoolkids streaming into Mammal Hall, the Hall of Human Origins, and rooms full of dinosaurs, reptiles, insects, gems, and minerals. I paused by an exhibit of award-winning nature photography, beautiful large-format color prints of wildlife from around the world. Birds featured prominently in the display, and one picture in particular seemed to catch everyone's attention.

It covered a wall where anyone turning the corner would meet it head-on—a vibrant four-by-six-foot shot of an Atlantic Puffin flying straight toward the camera. The bird filled the frame at an angle, its wings and bright-orange clown's feet splayed out as if it too had just veered around a corner. Somehow, the picture's every detail stood out in sharp focus, from the black rachis and charcoal vane of each tail feather to individual water droplets falling from the three silver minnows clamped in the bird's bill.

For several minutes I watched people turn the corner and encounter that giant puffin face-to-face—a mother and daughter, a young Japanese couple, a group of college-age women. They all reacted just as I had: a sudden intake of breath, then the leaning in for a closer look, the narrowing of the eyes, the careful examination. From surprise to query to wonder. Let the fascination begin.

Appendix A
An Illustrated Guide to Feathers

The following pages provide a visual reference for the feather types referred to throughout this book. These examples are by no means exhaustive, however—feathers are as diverse in form as they are in function. The drawings here show all the main feather varieties found in modern birds, but they really represent archetypes on a continuum. There's a wide range of possible fluffiness, for example, between down, semiplume, and contour feather. Some breeding plumes are so highly modified, they're hard to classify at all, and even bristles are far from uniform (see the "semibristle" pictured below). Nonetheless, this guide illustrates all the basic feathers one is likely to encounter and explains many of the terms used to describe them. The figures for feather growth, molting, and the developmental model of feather evolution are also reproduced again here for easy reference. All drawings by Nicholas Judson.

Flight Feather With its distinctive offset rachis, a flight feather can fit neatly between its neighbors to form a seamless wing or tail. But this shape also makes it remarkably aerodynamic on its own when those feathers are splayed. Flight feathers evolved from standard contour feathers and feature the same waterproof vanes with interlocking barbules. They can be brightly colored, elongated, or otherwise modified for breeding displays. In flightless birds, they often lose their aerodynamic qualities entirely and serve primarily for display, waterproofing, or other functions. Flight feathers are also known as remiges (singular: *remex*) for those on the wing and rectrices (singular: *rectrix*) when located on the tail. Remiges are further divided into primaries and secondaries, based on their position along the wing.

FLIGHT FEATHER

Contour Feather Contour feathers form the bulk of a bird's visible plumage. Their vanes are symmetrically arranged around the rachis, and the barbules interlock to form a smooth, waterproof whole. They often have an afterfeather, a plumy appendage at the base of the vane that adds additional insulation. Contour feathers vary greatly in size and appearance, from the tiny iridescent head feathers of a hummingbird to a duck's long belly feathers to the broad back and flank feathers on an eagle. Colors, breeding adaptations, and other variations abound—a peacock's display, the ear tufts of an owl, and the spongelike belly plumes of a sandgrouse are all modified contour feathers.

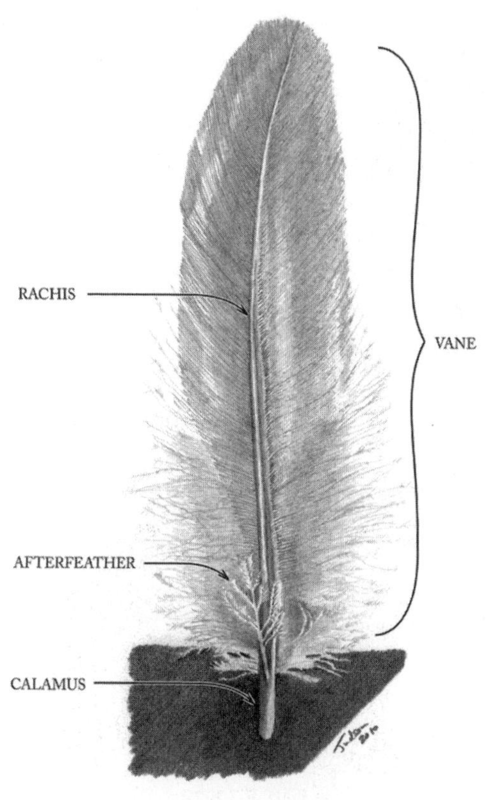

CONTOUR FEATHER

Semiplume Semiplumes fill the space in form and function between contour feathers and down. They have a distinct rachis, but their plumy barbs do not interlock to form a closed vane. They fill out the body plumage and add insulation, and in some cases the visible tips provide color. Semiplumes can also be modified for display. The Great Egret's elegant breeding plumes, once so desirable for ladies' hats, are simply elongated semiplumes.

SEMIPLUME

Down Feather A classic down feather has no rachis. Its barbs emerge from the rim of the calamus in a loose, springy tuft that gives them their remarkable insulative qualities. Practically speaking, the term *down* is often applied to any insulative feather, and there are a wide variety of downlike plumes with short or partial rachises. Many down feathers have elongated barbules, adding loft and increasing their ability to trap and hold air.

DOWN

Bristles Bristles are essentially a stiff, barbless or lightly barbed rachis. They serve a sensory or protective function in some cases and are often found on the face, feet, or other areas of bare skin. (Those pictured here are from the face and feet of a Barn Owl.) In birds that catch insects from the air, long bristles often help direct prey toward their mouths. They vary in form—those with enough barbs to begin resembling a standard vane are known as semibristles.

Filoplume Unlike most other feathers, filoplumes lack muscles within their follicles and cannot be adjusted or moved independently. Their role is sensory, providing the bird with information on the movement and condition of other feathers nearby. Clusters of filoplumes usually surround the base of each flight feather. They act like telltales on a sail, giving instant data on wind speed and feather position and helping the bird make fine adjustments during flight. In rare cases, elongated filoplumes serve a function in breeding displays, elongating so that their tufted tips emerge dramatically from the surrounding body feathers.

FILOPLUME

Developmental Theory of Feather Evolution The developmental theory proposes a series of cumulative evolutionary steps leading to modern feathers: an unbranched quill (Stage I), simple filaments (Stage II), filaments centered on a rachis (Stage III), interlocking barbules and pennaceous vanes (Stage IV), and asymmetrical flight feathers (Stage V).

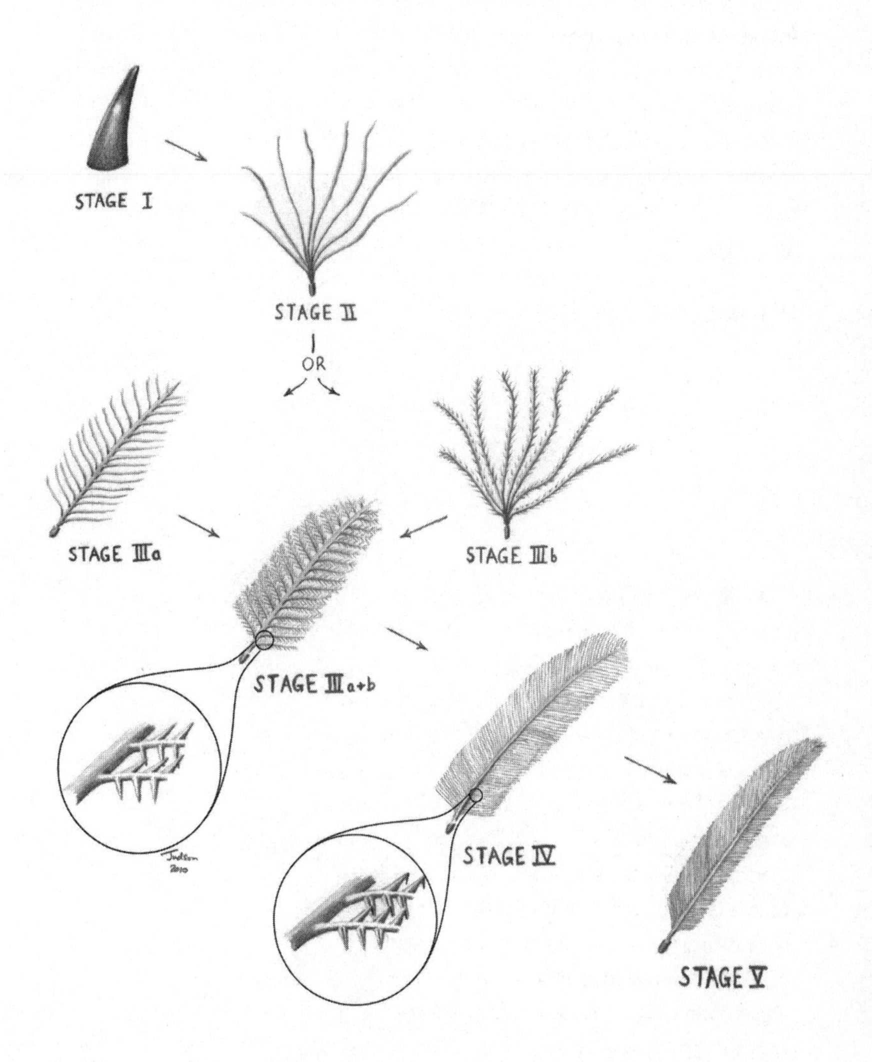

Feather Growth and Molt In the first illustration, the calamus of a mature feather nestles tightly inside its follicle, disconnected from the blood flow of the live dermal tissue below. When the molt begins (second illustration), cell activity in the follicle collar starts producing the barbs and rachis of a new feather, nourished by live tissue that extends up inside the arc of the developing barbs. The old feather is pushed out, and the new one grows to take its place, barbs unfurling to form a vane as they emerge from the follicle and a temporary protective sheath (third illustration). When the vane is complete, the follicle collar produces a solid tube of keratin, the calamus, that forms the base of the feather. Growth then stops, the live tissue recedes, and we return to the scene in the first drawing: a mature feather of dead keratin cut off from the live tissue below. The growth detail shows how barbs form in a helical fashion around the rim of the follicle collar before fusing with the solid rachis and proceeding upward.

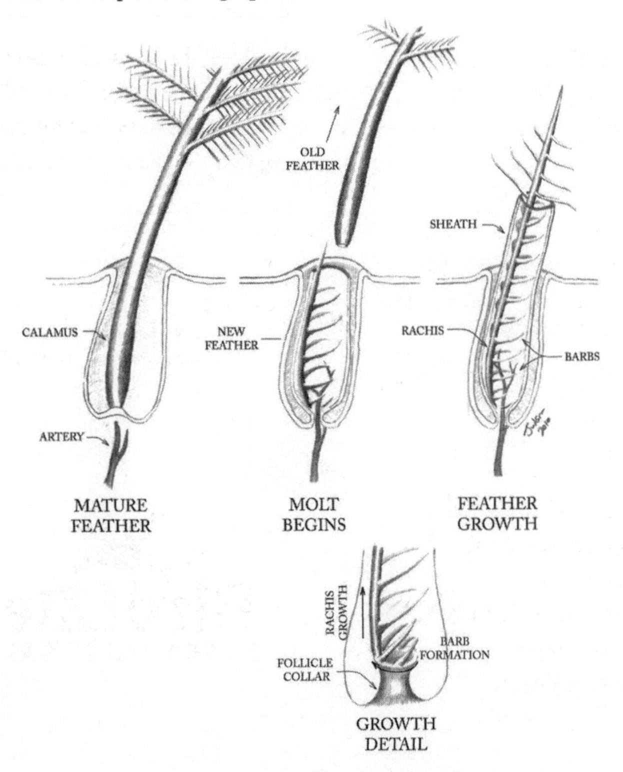

Appendix B

Feathers and Conservation

A portion of the proceeds from this book will be donated to help repay our debt of wonder, protecting, and preserving birds (and their feathers) in the wild. To support bird conservation efforts directly, consider making a donation to one of the following four estimable organizations.

National Audubon Society
225 Varick Street, 7th Floor
New York, NY 10014
Phone: (212) 979-3000
www.audubon.org

BirdLife International
Wellbrook Court
Girton Road
Cambridge CB3 0NA, UK
Phone: +44 (0)1223 277 318
www.birdlife.org

Cornell Lab of Ornithology
159 Sapsucker Woods Road
Ithaca, NY 14850
Phone: (800) 843-2473
www.birds.cornell.edu

American Bird Conservancy
4249 Loudoun Avenue
The Plains, VA 20198-2237
Phone: (888) 247–3624
www.abcbirds.org

Notes

Feathers are an enormous topic. Research for this book has spanned subjects as diverse as ornithology, aeronautical engineering, paleontology, mythology, penmanship, and the history of sports. I've read textbooks, explorers' memoirs, fashion magazines, and old newspapers, not to mention scientific journals ranging from *Polar Biology* to the *Australian Journal of Anthropology* to the *Journal of Colloid and Interface Science*. In this section I note additional points of interest and suggest a few key references for each chapter—books and articles that will give the interested reader a deeper look into the subjects that make feathers so fascinating. See the Bibliography for complete source information, including full author name and publication information.

Preface

xiv **As ornithologist Frank Gill observed:** Frank Gill's textbook, *Ornithology* (2007), provides an excellent introduction to feathers—their evolution, biology, and overall importance to birds. The text is clear, well written, and highly recommended.

Introduction

8 **"Each Waorani has a body and two souls":** Davis 1996, 271–272.

Chapter One

15 **It was the first full specimen:** Two books (at least) have been devoted to *Archaeopteryx* in recent years, and both provided important background for this chapter (see Shipman 1998 and Chambers 2002). Thomas Huxley's original description of *Archaeopteryx* also makes very worthwhile reading (1868).

Chapter Two

34 **With its head cocked, eyes open, and stumpy tail turned up:** Winter Wrens, Golden-crowned Kinglets, and indeed most other birds in North America are protected under the Migratory Bird Act and other laws. Collection or possession of any parts of these species is prohibited without permits from the U.S. Fish and Wildlife Service and relevant state agencies.

37 **a theory put forth by Dr. Richard Prum:** The definitive papers on feather evolution are by Rick Prum and Alan Brush (see Prum 1999 and Prum and Brush 2002).

40 **"I'm not too keen on neutral evolution":** For an articulate BAND viewpoint, read Alan Feduccia's *The Origin and Evolution of Birds* (1999).

42 **an exchange student had walked to the front of the room:** One of a group of talented young paleontologists, Dr. Zhonghe Zhou went on to direct China's Institute of Vertebrate Paleontology and Paleoanthropology and has participated in the discovery of numerous early birds and feathered dinosaurs.

Chapter Three

45 **where bands of ancient shales alternated with basalt:** Geologists use the term *formation* for a continuous series of rocks with no breaks in their chronology. The Yixian Formation includes a variety of sedimentary rocks interspersed with layers of basalt, reflecting long periods of deposition interrupted by volcanic activity. It reaches a depth of more than one kilometer in places.

45 **the fine-grained Yixian rocks preserve amazing levels of detail:** There are few books on the Yixian fossils, but see Mark Norell's *Unearthing the Dragon* (2006). Scientific papers on feathered dinosaurs continue to be published at a rapid pace, and there will undoubtedly be new discoveries before this book goes to press. See the Xu Xing papers listed in the Bibliography, and keep an eye on the science column in your local newspaper!

48 **remnants of those feathers might have truly fossilized:** When bones fossilize, they are typically mineralized directly, a molecule-by-molecule replacement of the original material. For soft tissues like feathers, however, it's usually the by-products of their anaerobic decomposition that are preserved.

54 **Every dog in the world:** Dogs provide another important evolutionary lesson. If a domesticated wolf can lead to such variety in only a few thousand years, think of the evolutionary possibilities when that time frame expands to the 160 million years that dinosaurs roamed the earth, or the even longer period since the first feathers evolved.

Chapter Four

61 **a young Thin-billed Prion:** Since at least 1848, when famed naturalist and illustrator John Gould published the *Birds of Australia*, the word *prion* has referred to these dainty krill-feeding members of the family Procellariidae. In the 1980s, biochemist and future Nobel laureate Stanley Prusiner co-opted the term *prion* to describe the malformed proteins (from "*pro*teinaceous" and "*in*fectious") responsible for mad cow disease and related maladies. This ugly latter usage should in no way prejudice us against a beautiful group of seabirds!

62 **His illustrated field guide, *Seabirds*:** Never one to rest on his laurels, Peter tells me he is currently "mid-way through a 7 year project painting over 5,000 images for a new Seabirds Handbook." Its publication will be an event.

65 **Muttonbirders succeed through a simple but profound understanding:** For more background on muttonbirding, take a look at A. Anderson (1996).

67 **People don't generally eat feathers:** One notable exception to this rule occurs in Latin American cockfighting, where tradition has it that chewing a feather from your opponent's bird will bring good luck and success to your rooster.

70 **resting for months or even a full year:** An interesting exception to the molting pattern, powderdown feathers grow continually throughout a bird's lifetime. Scattered throughout the plumage, their fine tips disintegrate steadily into a "powder" of tiny keratin bits that dust the feathers and may confer waterproofing or other qualities. See the discussion in Chapter 13 for more details.

70 **Its ultimate structure:** The absolute classic book on feather growth and structure is by Alfred Lucas and Peter Stettenheim (1972). This is the book that led Rick Prum to his epiphany. The text is rather technical, but there are a lot of great photos and illustrations.

73 **Young birds acquire them in the nest:** Birds typically host a community of up to twelve different feather lice. Because they are passed from parents to offspring in the nest, or among adults at communal roosts, the lice associated with a particular bird species rarely come into contact with those from other birds. Since Thin-billed Prions, for example, do not rub shoulders with woodpeckers or crows, neither do their feather lice mingle and interbreed. The lice have therefore evolved in tandem with their bird hosts, developing a diversity that mirrors bird diversity almost exactly. (The exceptions to this rule include falcons, skuas, and other bird predators and scavengers, who may encounter a variety of feather lice while plucking their prey. The genetic makeup of their parasites, not surprisingly, is more cosmopolitan.)

74 **birds have developed elaborate ways to combat them:** Feather parasites are becoming a hot topic in ornithology. Though still equivocal, some studies suggest that birds take advantage of the antibacterial properties found in certain ant, snail, and fruit chemicals to help preserve their plumes. Sunbathing and dusting may also contribute to feather maintenance, and some preen oils appear to inhibit bacterial growth. It's possible that parasites helped lead to the evolution of molting itself, as well as the development of feather colors, since dark, high-melanin feathers appear more resistant to damage.

Chapter Five

86 **This habit probably developed:** Recent experiments confirm this theory. When presented with a range of predators, from ferrets to falcons to owls, captive Black-capped Chickadees made distinctly different alarm calls to reflect the size and potential threat of each predator. Small bird-eating specialists like the Northern Pygmy Owl merited long, drawn-out warnings, while a Rough-legged Hawk (which eats mammals) received a short, dismissive phrase. Other chickadees, as well as Red-breasted Nuthatches, responded accordingly to the different calls. They based their behavior on the threat level encoded in the warning, a fascinating demonstration of communication within and between species at the heart of mixed winter flocks (see Templeton, Greene, and Davis 2005 and Templeton and Greene 2007).

88 **a fir branch covered by a thin layer of snow:** Snow and ice can provide a fair bit of protection from the elements (think igloos). Ruffed Grouse and other game birds of northern climes will often fly pell-mell into the powder at dusk, burying themselves completely in snug little snow caves.

88 **the incredible insulative quality of feathers:** Bernd Heinrich's *Winter World* (2003b) discusses feathers and a wide range of other animal adaptations to winter survival. Peter Marchand's *Life in the Cold* (1996) is also good.

90 **they manipulate the feathers to trap or release heat:** Smithsonian Institute researcher Dr. Carla Dove has documented incredible adaptive intricacies in the barbules of down feathers. Dabbling ducks, for example, have large triangular nodes between each cell that help trap more air and maximize each feather's insulative potential. Diving ducks live in the same cold water, but they can't afford a plumage that holds so much air—it would make them too buoyant to dive and feed below the surface. The nodes on their barbules are distinctly smaller, and their down is less efficient. They presumably make up the difference with increased body fat or other metabolic changes, but that question has yet to be studied.

90 **searching tirelessly for their roost sites:** Just where kinglets spend the night remained something of a mystery until quite recently, when Bernd finally succeeded in photographing four of them huddled together on a bushy white pine branch. It confirms that they do indeed pass those frigid nights in the open, or under a thin layer of snow.

90 **an astonishing 140 degrees Fahrenheit (78 degrees Celsius):** Kinglets maintain a body temperature of 111 degrees Fahrenheit (44 degrees Celsius), and there's no indication that they go into torpor, even on nights as cold as –29 degrees Fahrenheit (–34 degrees Celsius).

92 **Pacific Coast Feather makes and sells millions of pillows and comforters every year:** The history of the Pacific Coast Feather Company, as well as a lot of insight into the global trade, is told in *A Good Night's Sleep* (Roush and Beck 2006).

93 **consumption of goose and duck meat:** More than 99 percent of the world's feather and down comes as a by-product of the meat trade. A tiny market for live-plucked down persists, but most processors and retailers eschew this practice as both uneconomical and cruel to the birds.

98 **a full storm suit with matching mukluks and mittens:** The clothing weighed for this example was sewn in the 1950s by a member of the Inupiaq Eskimo community of Point Hope, Alaska. The outfit consisted of a man's inner parka (five pounds), outer parka (four pounds), trousers (five pounds), and knee-length boots (four pounds). All were made from caribou skins, with wolverine hood linings and bearded-seal boot soles. (Measurements most generously provided by Angela Linn of the University of Alaska's Museum of the North.)

99 **a sodden mess that loses much of its insulative value:** Feather mixtures, however, take advantage of the natural water resistance in contour feathers and perform much better wet than down alone.

Chapter Six

102 **Mammalian athletes also help cool themselves by sweating:** Though all mammals have sweat glands (from which the milk-producing

mammary glands evolved), many species lack enough of them for effective temperature regulation. Primates, camels, and horses are notoriously good at sweating, while dogs, felines, rodents, rabbits, and many other groups rely more on panting and other adaptations.

103 **birds can't afford to heat up:** Notable exceptions include ostriches, as well as various doves, quail, sandgrouse, and other birds of hot, arid climates. These species regularly use *hyper*thermia as a temperature regulation strategy. They allow their body temperatures to increase to near-lethal levels so that they can simply radiate excess heat rather than risk the water loss inherent in panting.

109 **tracts separated by large areas of bare skin called *apteria*:** Ornithologists call these feather tracts *pterylae* (thus *apteria* for bare patches). Most species have eight main tracts and dozens of subgroups arranged into distinctive patterns. Before the advent of DNA analysis, ornithologists studied these patterns to unravel close evolutionary relationships among birds, and "pterylosis" remains an important tool for bird taxonomy.

109 **plenty of other options for dispersing heat:** Penguins generate the most heat while swimming, using their stubby wings to "fly" gracefully below the surface. This type of locomotion works only where the water is cold enough to dispel the heat from pectoral muscles. Auks do it in the northern oceans and penguins in southern, but the habit is unknown in tropical seas. Marine mammals like whales and dolphins can swim normally throughout the tropics, but a bird's feathered insulation makes flapping in warm water physiologically impossible.

111 **their unique respiratory system takes evaporative cooling:** Avian respiration and cooling are well described in Gill 2007.

112 **a panting bird can take hundreds of breaths per minute:** Some species (particularly herons, pelicans, owls, game birds, and goatsuckers) increase heat loss further by rapidly vibrating bones and membranes in their upper throat, a process called "gular fluttering."

112 **breathing helps keep an active bird from overheating:** Water loss is an unavoidable side effect of evaporation, but birds appear to avoid getting dehydrated by keeping most flight times short and by ascending to high, cool altitudes for long migratory journeys.

113 **bat bodies retain a suite of much simpler cooling options:** See Reeder and Cowles 1951 for a classic paper on bat thermoregulation and Hedenström, Johansson, and Spedding 2009 for a technical comparison of bats and birds.

Chapter Seven

118 **All three were Silver-laced Wyandottes:** The later addition of two Rhode Island Reds added a splash of color and brought our flock up to a total of five, before a hungry Bald Eagle reduced the count to four, where things stabilized.

118 **one of ornithology's most divisive questions:** Literature on the ground-up–tree-down debate could fill a bookshelf, but it's probably best to start with the classic viewpoints: Ostrom 1979 and Feduccia 2002.

123 **Even bats, the most nimble nonavian fliers:** *Nimble* hardly does bats justice. Recent studies of bat flight show novel thrust and lift mechanisms that give them incredible maneuverability, particularly at low speeds. A bat in pursuit of its insect prey can execute 180-degree turns in the distance of half a wingspan.

123 **feathers seem grossly overqualified for the job:** A related and equally significant problem is that the vast majority of "tree-down" examples involve gliding, rather than the powered flapping flight that birds use. Not one of dozens of extant gliders shows any evolutionary inclination toward flapping. As one ornithologist I spoke to put it, "Gliding is a perfect adaptation for gliding, but it's a dead end for flight."

127 **help them scramble up otherwise impossible slopes:** In the wild, Chukar Partridges inhabit dry, rocky grass and shrublands, where they're vulnerable to a wide range of predators. They often take refuge in caves or crevices between bouts of foraging or when threatened. Using WAIR allows even the youngest birds to ascend cliffs and rocks to safety, an advantage with immediate impacts on survival.

127 **Ken called the technique WAIR:** For more details on the wing-assisted incline running theory, see Dial 2003 and Dial, Randall, and Dial 2006.

Chapter Eight

132 **he dropped the hammer and the feather together:** There is a wonderful Web site with detailed log entries and interviews about the Apollo 15 mission (see E. Jones 1996). It also features a link to the video showing the feather and hammer experiment.

135 **once both he and the falcon are airborne:** For information on Ken Franklin's work with falcons, see Harpole 2005 and seek out the National Geographic movie *Terminal Velocity*.

Chapter Nine

141 **the *mimicking* of *biological* structures, behaviors, and processes:** Though it has very little to do with feathers, Robert Allen's *Bulletproof Feathers* (2010) gives a beautifully illustrated introduction to modern efforts in biomimicry.

142 **it was a short leap to fletching:** Although there may be an intuitive connection between bird flight and archery, properly fletching an arrow is anything but simple. Ishi, the last member of California's Yahi people, emerged from the wild in 1911 with the expert knowledge of a culture that had depended on their archery skills for sustenance and defense. His doctor, friend, and bow-hunting apprentice Saxton Pope observed, "Many kinds of feathers were used by Ishi on his arrows—eagle, hawk, owl, buzzard, wild goose, heron, quail, pigeon, flicker, turkey, bluejay. . . . Like the best archers he put three feathers from the same wing on each arrow." Ishi's process included intricate cutting, trimming, and notching, followed by adhesion with strips of chewed deer sinew. The size and shape of the fletching matched specific purposes, from narrow three-inch vanes on small hunting arrows to war shafts with feathers "the full length of a hawk's pinions—almost a foot." Such precise craftsmanship made fletching one of the earliest specialty trades, and ancient armies required hundreds or even thousands of dedicated arrow makers. At its peak, for example, Genghis Khan's famous light cavalry rode into each battle equipped with more than nine million handmade arrows.

144 **a grand tale that's been well told elsewhere:** Of the many books on this topic, James Tobin's *To Conquer the Air* is particularly good.

145 **When air meets the front of a bird wing, it has a choice:** In spite of the fact that every bird (or airplane) aloft relies on these principles to fly, the relative contribution of airfoil shape, angle, air pressure, and other factors remains contentious. Flowing air creates complex patterns of pressure, currents, and vortices, and although engineers can calculate an airfoil's performance precisely, the actual process retains some degree of mystery. For an excellent explanation of the current thinking, see D. Anderson and Eberhardt 2001.

146 **notions about birds and airfoils didn't emerge again until the late 1800s:** For more on the history of human flight, I highly recommend Octave Chanute's classic *Progress in Flying Machines* (1894) as well as Lilienthal 2001 and Tobin 2003.

149 **aircraft designers had essentially perfected the form:** D. Anderson and Eberhardt 2001 provides excellent explanations of the mechanics of flight.

152 **rough surfaces can help reduce drag:** Sharkskin behaves this way in water, echoing Leonardo's observations that air and water respond to the same principles. This effect led swimwear companies to design scaly full-body suits that reduced drag for competitive swimmers by up to 5 percent. Swimmers wearing "sharkskin" suits broke more than 250 world records in less than two years before the material was finally banned from competition.

154 **extreme sports may offer a glimpse of this feeling:** Or not. My curiosity once got the best of reason, and I tried bungee jumping, diving from a high bridge with nothing but an elastic rope tied around my ankles. Jumping off bridges, it turns out, is nothing like flying. It's like falling.

154 **"I sometimes think that the desire to fly":** McFarland 1953.

Chapter Ten

159 **other naturalists have suffered the exact opposite problem:** Alfred Russel Wallace's *Malay Archipelago* (1869) is a wonderful introduction to birds of paradise and the natural history of modern-day Indonesia. Frith and Beehler's *The Birds of Paradise* (1998) is an excellent reference, with great illustrations by William T. Cooper.

160 **Reading his description of their "dancing parties":** Wallace 1869.

163 **Darwin's second major contribution to evolutionary theory:** Darwin 1871 provides the foundation for sexual selection, but see Johnsgard 1994 or Hill and McGraw 2006b for modern interpretations of the theory.

164 **evolutionary biologists now distinguish between two basic forms of sexual selection:** Though intra- and intersexual selection are useful generalizations, each includes subtleties, and the barrier between them is far from fixed. Females in many intrasexual systems can make choices based on visual cues or a male's performance in combat. Similarly, many highly adorned male birds have been known to fight to the death defending breeding rights or territories.

165 **feathers and arguably birds themselves would never have become so diverse:** Many experts now lean toward display and mate choice as perhaps the strongest factors in the early development and diversification of feathers. By the time of the first beaked bird, *Confuciusornis* (another Yixian Formation find), sexual dimorphism was well established. Whole flocks of these crow-size fliers have been unearthed, with males sporting wildly elongated tail feathers indistinguishable from those of various modern widowbirds, flycatchers, or birds of paradise.

167 **Wallace countered these fables with careful observation:** There is an irony in using Wallace and birds of paradise as the prime example of Darwinian sexual selection, since Wallace himself never cottoned to the theory. He believed that natural selection through male competition accounted for variations in plumage and other features and thought that Darwin put too much emphasis on female choice. Later studies favor Darwin's interpretation.

Chapter Eleven

176 **The global plume trade peaked in the years before World War I:** Good references on different aspects of the plume trade include Stein 2008, Swadling 1996, and Price 1999.

176 **one fashion craze drove the entire industry:** By one count, the millinery trade in 1900 employed more than one in every three hundred American workers. In modern terms, that would equal

well over a half-million people, more than the active memberships of the United Auto Workers, the Longshoremen, the United Farm Workers, the Association of Flight Attendants, and the Writers Guild of America combined.

176 **feathers had epitomized high style:** Feathered head wear developed independently in numerous prehistoric cultures but formally entered Western tradition by way of Persia, where soldiers added a "feather to their cap" to commemorate battlefield kills. Feathers still feature prominently in military dress uniforms around the world, from Scotland's Black Watch (red hackle) to Italy's Bersaglieri (grouse) and the Swiss Guard (ostrich).

179 **"Whoever wears an ostrich plume":** Hayden 1913.

183 **"I am sorry I cannot give the full particulars":** From a letter to George Aschman, August 9, 1944.

185 **"During the two years I was in England":** Smith various.

186 **fall from the rachis in long, extravagant waves:** The rachides of ostrich plumes divide their wide vanes into perfectly equal halves, a quality that led ancient Egyptians to revere them as powerful symbols of truth, law, and morality. Ostrich plumes appeared commonly in hieroglyphics, decorated the crown of Osiris, and were the talismans of Maat, goddess of justice.

188 **a conservation ethic that still resonates in the modern environmental movement:** Frank Chapman's autobiography (1933) is a good read and provides insight into the development of the bird-conservation movement.

Chapter Twelve

198 **a range of other functions also added pressure for the evolution of color:** There is a wonderful new National Geographic book on bird coloration (Hill 2010), which describes the evolution, natural history, and physics of feather colors in very readable prose (with excellent color photos). For the science behind this book, see Hill and McGraw 2006a, 2006b.

198 **the melanin tinting a sparrow feather:** In addition to color, the presence of melanins can toughen feathers, making them more resistant to physical wear and rot. This explains why many birds in wet cli-

mates are darker than their relatives elsewhere (for bacterial resistance), why birds that move through abrasive vegetation are dark (e.g., rails, crakes), and why flight feathers exposed to heavy wear are often dark (e.g., the wing tips of gulls and buteos). A recent study found that the brilliant red, orange, and yellow pigments unique to parrot feathers are also resistant to bacteria, a useful adaptation in their humid rain forest environment. (See Burtt et al. 2010.)

201 **the result is an unparalleled diversity of colors and effects:** Although the common description of birds as "living jewels" may sound trite, it's entirely accurate in the case of the Opal-crowned Manakin. Crystalline structures within its feather barbs mimic those of an opal almost exactly and scatter light in the same way. For all intents and purposes, this little Amazonian songbird is flying around with a gemstone on its head.

203 **Feather money, or *tevau*:** In the local tongue, *tevau* is simply a generic word for money. The people of Santa Cruz and neighboring islands referred to feather coils by at least eleven different names, depending on their length, condition, age, and the quality of the feathers. But with local spellings like *lrdq*, those words are usually avoided in favor of the generic term.

204 **featherwork continues in the art and adornment of many cultures:** To learn more about featherwork, seek out the well-written and beautifully illustrated volumes by Reid (2005) and Reina and Kensinger (1991).

205 **"I am forced to abstain":** B. Díaz del Castillo [1570] 1956.

206 **even common citizens kept colorful songbirds as pets:** This tradition persists in many parts of Latin America. Fiercely contested songbird competitions transform the central square of Paramaribo, Suriname, every Sunday afternoon, briefly filling the sleepy capital with the warbles and trills of the rain forest.

206 **outlawing traditional practices, including featherwork:** Among few exceptions to this rule was the Franciscan missionary and historian Bernardino de Sahagún (1499–1590), who encouraged surviving feather artists to use their talents for Christian themes. Several bishops' miters, triptychs, and other ecclesiastical pieces survive, resembling the finest Renaissance paintings but wrought in hummingbird feathers.

207 **with tithes paid in birds and feathers from all corners of their domain:** Feathers even figured in the causes and spoils of Incan warfare. The Inca conquest of the Cuyo people stemmed in part from the Cuyos' refusal to trade "certain birds found in that land." Following the Inca victory, a thousand cages of Cuyo birds were sent to the emperor in tribute.

208 **retained their vibrant coloration for more than a thousand years:** The Inca inherited a long-standing tradition of featherwork in Peru, and surviving artifacts date to a number of earlier cultures, including the Nazca (AD 100–600), the Huari (AD 600–1000), and the Chimú (AD 900–1500).

Chapter Thirteen

218 **structure is the key to waterproofing:** Several fascinating scientific articles provided information on waterproofing (see Bormashenko et al. 2007, Ortega-Jiminez and Alvarez-Borrego 2010, and Yang, Xu, and Zhang 2006).

219 **before their preen glands have even started producing oil:** Ornithologists used to assume that the mother used her own oils to preen her chicks in preparation for their first swim, since hatchery-reared birds would invariably get wet and drown. Now it seems that the hatchery chicks suffered only from lack of sanitation: amniotic residues left over from hatching made their down more permeable. Once cleaned, the domestic chicks were perfectly able to stay dry in water for hours without preen oils. Exactly when and how a mother Mallard cleans her chicks (or whether they do it themselves) remains a mystery.

221 **They benefit from the negative buoyancy of soaking:** All diving birds face the same buoyancy conundrum. Their waterproofing maintains a lifesaving layer of warm, dry air against their skin and down, but that very air makes it harder to dive and stay beneath the surface. Studies show that the structure of down feathers in diving birds holds less air (and is therefore less insulative) than those of dabblers (which is why the best feather beds are filled with down from geese and not cormorants or mergansers). Diving

birds presumably compensate with increased body fat or some metabolic difference, but this has yet to be studied.

222 **a peculiar sandgrouse quirk:** This adaptation raises all kinds of evolutionary questions. As Darwin noted, differences in plumage generally derive from sexual selection, and many sandgrouse species are known for the male's "breast display" posture during mating rituals. Could female sandgrouse choose spongy breasts the same way that birds of paradise prefer bright colors, long tails, and elaborate dances? Are the coil-barbed feathers present only in breeding plumage? What is the cost to the male in terms of insulation or vulnerability to rainstorms? These questions have yet to be examined—in spite of their fascinating idiosyncrasies (and great beauty), sandgrouse have received almost no research attention in more than forty years.

223 **Fly-fishing as sport and compulsion:** There are a host of great books on fly-fishing and fly-tying, many with great illustrations of feathered flies. As a start, look for Walton 1896, Kelson 1895, and Schmookler and Sils 1999.

223 **"They fasten red (crimson red) wool around a hook":** Translation from the Greek, as quoted in Radcliffe 1921.

226 **his wealth of detailed knowledge:** I asked John to focus on feathers, but fly-tying materials don't stop there. He rattled off a long list of things he's used over the years to get a particular effect, from yarn and wool to deer fur, colored beads, and even polar-bear hair. "The big thing these days," he said, rolling his eyes, "is the fur off the bottom of the pads on the back feet of a snowshoe hare!" I looked it up online when I got home—he wasn't kidding!

229 **vintage collections now command surprising sums at auction:** There is a dark side to this hobby, however. Many of the feathers used by nineteenth-century tiers came from birds that are now rare or even endangered in the wild. Yet some collectors and makers insist on re-creating those patterns *exactly* as they first appeared, regardless of the consequences. This demand creates a lucrative niche market for their feathers, adding to the stress on populations already suffering from habitat loss, overhunting, and other pressures.

Chapter Fourteen

236 **the first definite reference to the use of a quill:** One of the best books on the history of writing is Carvalho's *Forty Centuries of Ink* (1904). Finlay (1990) is also excellent.

236 **Production of quill pens peaked in the early nineteenth century:** For a good description of the quill trade, see "History of Writing Materials" 1838.

237 **common scene of a schoolboy timidly addressing his writing master:** Though the articles in *Household Words* were unsigned, the "urchin" in this scene is strongly reminiscent of little Oliver Twist, whose famous workhouse request of, "Please, sir, I want some more," first appeared in print only twelve years earlier.

245 **They remained a viable competitor to wooden picks:** The curious history of quill toothpicks deserves a chapter of its own. Happily, Henry Petroski has already accomplished that task admirably in his book, *The Toothpick* (2007, chap. 4).

245 **featherbone powder puffs may have faded away:** Patented in 1883 by Edward K. Warren, "featherbone" transformed turkey quills discarded by the feather-duster industry into a low-cost alternative to the whale baleen used in corsets, bustles, bust extenders, and other fashion essentials of the day (as well as powder puffs). Mr. Warren made a fortune.

245 **engineers and entrepreneurs continue to find surprising new uses for the plumage of birds:** Information on the various industrial uses of feathers is scattered throughout the scientific literature and patent-office databases.

Chapter Fifteen

251 **it's easy to imagine natural selection favoring baldness:** If you were to run your hand over a vulture's knobby head, you would probably find a few bristles or short quills topped with bits of fuzz. I've tried this with museum specimens, and it turns out that vultures aren't entirely bald. Some feathers have been lost, but others remain, either complete or reduced to simpler forms. The degree of

feather loss is related to the evolutionary trade-off between hygienic advantage and the risk of heat loss. Bare heads stay clean, but they lack the insulating and protective qualities that come with feathers. The balance tips in favor of baldness only for the messiest eaters, those species that regularly feed inside the body cavities of big animals. At our carcasses in Kenya, the griffons and Nubian Vultures had bare heads, while the diminutive Egyptian Vulture was almost fully feathered, its baldness limited to a yellow patch around the face. Unable to compete directly with its larger cousins, the Egyptian Vulture focused on cleaning up after they were through, using its small bill to glean meat and sinew from joints and crevices too narrow for the larger birds. Its head and neck rarely become fully soiled, and the feathers persist.

259 **rubbing body parts together in what scientists call *stridulation*:** Stridulation is extremely uncommon in all vertebrates. Simplified forms are known from snakes such as the saw-scaled viper, which rasps its scales together with a sizzling sound during threat displays. Fish, too, have been known to stridulate with gill bones or spines, but the technique is most common in insects, whose hard exoskeletons, membranous wings, and rapid-fire musculature are best suited to it.

259 **Like crickets, grasshoppers use a pick-and-file system to stridulate:** The pick and file alone does not produce the cricket's distinctive chirr, but it's just at the right frequency to cause the wing membrane to resonate and broadcast the tone. This system inspired Kim to take her research further, and she's now shown that the flight-feather shafts throughout the manakin's wing resonate with the "ting" and that the potential to vibrate at that frequency seems inherent in all feathers. The Club-winged Manakin's adaptation is unique, but it takes advantage of a sonic characteristic intrinsic in feather structure.

Conclusion

265 **a collection of more than 650,000 stuffed birds:** The third-largest bird collection in the world, the Smithsonian's National Collection

includes skins, nests, eggs, skeletons, and tissue samples used for a wide range of ornithology research by resident scientists as well as visiting scholars.

266 **Though bird (et al.) strikes remain their most common cases:** Since pilots record the elevation of each incident, the lab's massive bird-strike data set is beginning to change the way people think about flight and migration habits. They've had Griffon Vulture strikes at thirty-seven thousand feet, ducks at twenty-seven thousand feet, and shorebirds as high as twelve thousand feet. People used to think these were rare events, but it now appears that even songbirds regularly migrate at extreme elevations.

Appendix A

277 **Feather Growth and Molt:** This figure pictures the growth of a typical vaned feather. For down, bristles, filoplumes, and other nonvaned types, the process is very similar but varies in the arrangement of barbs and the presence or absence of a rachis.

Bibliography

Aiken, Charlotte Rankin. 1918. *The millinery department*. New York: Ronald Press.

Allen, Grant. 1879. Pleased with a feather. *Popular Science Monthly* 15: 366–376.

Allen, Robert, ed. 2010. *Bulletproof feathers: How science uses nature's secrets to design cutting-edge technology*. Chicago: University of Chicago Press.

Anderson, Atholl. 1996. Origins of *Procellariidae* hunting in the Southwest Pacific. *International Journal of Osteoarchaeology* 6: 403–410.

Anderson, David F., and Scott Eberhardt. 2001. *Understanding flight*. New York: McGraw-Hill.

Attenborough, David. 2009. Alfred Russel Wallace and the birds of paradise. Centenary Lecture, Bristol University, September 24, 2009.

Audubon, John James. 2008. *120 Audubon bird prints*. Mineola, NY: Dover Publications.

Baier, Stephen. 1977. Trans-Saharan trade and the Sahel: Damergu, 1870–1930. *Journal of African History* 18: 37–60.

Bakken, George S., Marilyn R. Banta, Clay M. Higginbotham, and Aaron J. Lynott. 2006. It's just ducky to be clean: The water repellency and water penetration resistance of swimming mallard *Anas platyrhynchos* ducklings. *Journal of Avian Biology* 37: 561–571.

Barbosa, A., S. Merino, J. J. Curevo, F. De Lope, and A. P. Moller. 2003. Feather damage of long tails in Barn Swallows *Hirundo rustica*. *Ardea* 91: 85–90.

Barney, Stephen A., W. J. Lewis, J. A. Beach, and Oliver Berghof, trans. 2006. *The etymologies of Isidore of Seville*. Cambridge: Cambridge University Press.

Barrett, Paul M. 2000. Evolutionary consequences of dating the Yixian Formation. *Trends in Ecology and Evolution* 15: 99–103.

Bartholomew, George A., Robert C. Lasiewski, and Eugene C. Crawford Jr. 1968. Patterns of panting and gular flutter in cormorants, pelicans, owls, and doves. *Condor* 70: 31–34.

Begbie, Harold. 1910. New thoughts on evolution: Views of Professor Alfred Russel Wallace. *Daily Chronicle* (London), November 3–4, 4.

Belloc, Hilaire. 1897. *More beasts for worse children*. London: Duckworth.

Bewick, Thomas. 2004. *Bewick's animal woodcuts*. Mineola, NY: Dover Publications.

Boerger, Brenda H. 2009. Trees of Santa Cruz Island and their metaphors. From "Proceedings of the Seventeenth Annual Symposium About Language and Society, Austin." *Texas Linguistic Forum* 53: 100–109.

Bonser, Richard H. C. 1995. Melanin and the abrasion resistance of feathers. *Condor* 97: 590–591.

Bonser, Richard H. C., and C. Dawson. 1999. The structural mechanical properties of down feathers and biomimicking natural insulation materials. *Journal of Materials Science Letters* 18: 1769–1770.

Borgia, Gerald. 1985. Bower quality, number of decorations, and mating success of male Satin Bowerbirds (*Ptilonorhynchus violaceus*): An experimental analysis. *Animal Behavior* 33: 266–271.

Bormashenko, Edward, Yelena Bormashenko, Tamir Stein, Gene Whyman, and Ester Bormashenko. 2007. Why do pigeon feathers repel water? Hydrophobicity of pennae, Cassie-Baxter wetting hypothesis and Cassie-Wenzel capillarity-induced wetting transition. *Journal of Colloid and Interface Science* 311: 212–216.

Bonshek, Elizabeth. 2009. A personal narrative of particular things: *Tevau* (feather money) from Santa Cruz, Solomon Islands. *Australian Journal of Anthropology* 20: 74–92.

Bostwick, Kimberly S. 2000. Mechanical sounds and evolutionary relationships of the Club-winged Manakin (*Machaeropterus deliciosus*). *Auk* 117: 465–478.

Bostwick, Kimberly S., Damian O. Elias, Andrew Mason, and Fernando Montealegre-Z. 2010. Resonating feathers produce courtship song. *Proceedings of the Royal Society B* 277: 835–841.

Bostwick, Kimberly S., and Richard O. Prum. 2003. High-speed video analysis of wing-snapping in two manakin clades (*Pipridae: Aves*). *Journal of Experimental Biology* 206: 3693–3706.

———. 2005. Courting bird sings with stridulating wing feathers. *Science* 309: 736.

Brigham, William T. 1918. *Additional notes on Hawaiian featherwork: Second supplement.* Memoirs of the Bernice Pauahi Bishop Museum, vol. 7, no. 1. Honolulu: Bishop Museum Press.

Bryant, David M. 1983. Heat stress in tropical birds: Behavioural thermoregulation during flight. *Ibis* 125: 313–323.

Burtt, Edward H., and Jann M. Ichida. 2004. Gloger's rule, feather-degrading bacteria, and color variation among Song Sparrows. *Condor* 106: 681–686.

Burtt, Edward H., Max R. Schroeder, Lauren A. Smith, Jenna E. Sroka, and Kevin J. McGraw. 2010. Colourful parrot feathers resist bacterial degradation. *Biology Letters* doi: 10.1098/rsbl.2010.0716.

Byron, Lord. 1809. *English bards and Scotch reviewers.* London: James Cawthorn.

Cade, Tom J., and Gordon L. Maclean. 1967. Transport of water by adult sandgrouse to their young. *Condor* 69: 323–343.

Calder, William A. 1968. Respiratory and heart rates of birds at rest. *Condor* 70: 358–365.

Canals, M., C. Átala, R. Olivares, F. Guajardo, D. Figueroa, P. Sabat, and M. Rosenmann. 2005. Functional and structural optimization of the respiratory system of the bat *Tadarida brasiliensis* (Chiroptera, Molossidae): Does the airway geometry matter? *Journal of Experimental Biology* 208: 3987–3995.

Carvalho, David N. 1904. *Forty centuries of ink.* New York: Banks Law.

Catry, Paulo, Ana Campos, Pedro Segurado, Monica Silva, and Ian Strange. 2003. Population census and nesting habitat selection of

Thin-billed Prion *Pachyptila belcheri* on New Island, Falkland Islands. *Polar Biology* 26: 202–207.

Chambers, Paul. 2002. *Bones of contention: The fossil that shook science.* London: John Murray.

Chanute, Octave. 1894. *Progress in flying machines.* New York: American Engineer and Railroad Journal.

Chapman, Frank Michler. 1886. Birds and bonnets. *Forest and Stream* 26, no. 6: 84.

———. 1908. *Camps and cruises of an ornithologist.* New York: D. Appleton.

———. 1933. *Autobiography of a bird-lover.* New York: D. Appleton–Century.

Chiappe, Luis M. 2007. *Glorified dinosaurs: The origin and early evolution of birds.* Hoboken, NJ: John Wiley and Sons.

Chiappe, Luis M., Jesús Marugán-Lobón, Shu'an Ji, and Zhonghe Zhou. 2008. Life history of a basal bird: Morphometrics of the Early Cretaceous Confuciusornis. *Biology Letters* 4: 719–723.

Christiansen, Per, and Niels Bonde. 2004. Body plumage in *Archaeopteryx*: A review, and new evidence from the Berlin specimen. *Comptes Rendus Palevol* 3: 99–118.

Clark, Christopher James, and Teresa J. Feo. 2008. The Anna's Hummingbird chirps with its tail: A new mechanism for sonation in birds. *Proceedings of the Royal Society* B 275: 955–962.

Clottes, Jean, ed. 2003. *Chauvet Cave: The art of earliest times.* Salt Lake City: University of Utah Press.

The commercial value of small things. 1891. *Chambers's Journal of Popular Literature, Science, and Arts* 68: 710–713.

Conard, Nicholas J., Maria Malina, and Susanne C. Müzel. 2009. New flutes document the earliest musical tradition in southwestern Germany. *Nature* 460: 737–740.

Coulson, David, and Alec Campbell. 2001. *African rock art.* New York: Harry N. Abrams.

Cowper, William. 1808. *Poems.* London: J. Johnson.

Darwin, Charles. 1859. *On the origin of species by means of natural selection; or, The preservation of favoured races in the struggle for life.* London: John Murray.

———. 1871. *The descent of man, and selection in relation to sex*. London: John Murray.

———. 1993. *The correspondence of Charles Darwin*. Vol. 8, *1860*. Cambridge: Cambridge University Press.

Davis, Wade. 1996. *One river*. New York: Simon and Schuster.

del Hoyo, Josep, Andrew Elliot, and Jordi Sargatal, eds. 1992. *Handbook of the birds of the world*. Vol. 1, *Ostrich to ducks*. Barcelona: Lynx Edicions.

Dial, Kenneth P. 2003. Wing-assisted incline running and the evolution of flight. *Science* 299: 402–404.

Dial, Kenneth P., Brandon G. Jackson, and Paolo Serge. 2008. A fundamental avian wing-stroke provides a new perspective on the evolution of flight. *Nature* 451: 985–989.

Dial, Kenneth P., R. J. Randall, and Terry R. Dial. 2006. What use is half a wing in the ecology and evolution of birds? *BioScience* 56: 437–445.

Diamond, A. W., and F. L. Filion, eds. 1987. *The value of birds*. ICBP Technical Publication, no. 6. Cambridge: International Council for Bird Preservation.

Díaz del Castillo, B. [1570] 1956. *The discovery and conquest of Mexico, 1517–1521*. Trans. A. Maudslay. New York: Farrar, Straus, and Cudahy.

Dickson, James G., ed. 1992. *The Wild Turkey: Biology and management*. Mechanicsburg, PA: Stackpole Books.

Dove, Carla, Marcy Heacker, and Bill Adair. 2004. In memorium: Roxie Collie Laybourne, 1910–2003. *Auk* 121: 1282–1285.

Drent, Rudolf Herman. 1972. Adaptive aspects of the physiology of incubation. In *Proceedings of the XVth International Ornithological Congress*, ed. K. H. Voous, 255–280. Leiden: E. J. Brill.

Dyck, J. 1985. The evolution of feathers. *Zoologica Scripta* 14: 137–154.

Eaton, Elon Howard. 1915. *Birds of New York*. Albany: New York State Museum.

Ehrlich, Paul R., David S. Dobkin, and Darryl Wheye. 1988. *The birder's handbook: A field guide to the natural history of North American birds*. New York: Simon and Schuster.

Favier, Julien, Antoine Dauptain, Davide Basso, and Allessandro Bottaro. 2009. Passive separation control using a self-adaptive hairy coating. *Journal of Fluid Mechanics* 627: 451–483.

Feduccia, Alan. 1999. *The origin and evolution of birds*. 2nd ed. New Haven: Yale University Press.

———. 2002. Birds are dinosaurs: Simple answer to a complex question. *Auk* 119: 1187–1201.

Feduccia, Alan, Theagarten Lingham-Soliar, and J. Richard Hinchliffe. 2005. Do feathered dinosaurs exist? Testing the hypothesis on neontological and paleontological evidence. *Journal of Morphology* 266: 125–166.

Feduccia, Alan, and Julie Nowicki. 2002. The hand of birds revealed by ostrich embryos. *Naturwissenschaften* 89: 391–393.

Finlay, Michael. 1990. *Western writing implements in the age of the quill pen*. Weterhal, England: Plains Books.

Ford, Horace Alfred. 1859. *Archery: Its theory and practice*. 2nd ed. London: J. Buchanan.

Frith, Clifford B., and Bruce M. Beehler. 1998. *The birds of paradise*. Oxford: Oxford University Press.

Frith, Clifford B., and William T. Cooper. 1996. Courtship display and mating of Victoria's Riflebird (*Ptiloris ictoriae*) with notes on the courtship displays of congeneric species. *Emu* 96: 102–113.

Gaston, Kevin J., and Tim Blackburn. 1997. How many birds are there? *Biodiversity and Conservation* 6: 615–625.

Gauthier, Jacques, and Lawrence F. Gall, eds. 2001. *New perspectives on the origin and early evolution of birds: Proceedings of the International Symposium in Honor of John H. Ostrom*. New Haven: Peabody Museum of Natural History, Yale University.

Gee, Henry. 1999. *In search of deep time*. New York: Free Press.

George, Brian R., Anne Bockarie, Holly McBride, Davi Hoppy, and Alison Scutti. 2003. Utilization of turkey feather fibers in nonwoven erosion control fabrics. *International Nonwovens Journal* 12: 45–52.

Gill, Frank B. 2007. *Ornithology*. 3rd ed. New York: W. H. Freeman.

Gill, Frank B., and D. Donsker, eds. 2010. IOC world bird names (version 2.5). http://www.worldbirdnames.org/. Accessed August 6, 2010.

Gleeson, Mike. 1985. Analysis of respiratory pattern during panting in fowl, *Gallus domesticus*. *Journal of Experimental Biology* 116: 487–491.

Godwin, Malcolm. 1990. *Angels, an endangered species*. New York: Simon and Schuster.

Gremillet, David, Christophe Chauvin, Rory P. Wilson, Yvon Le Maho, and Sarah Wanless. 2005. Unusual feather structure allows partial plumage wettability in diving great cormorants *Phalacrocorax carbo*. *Journal of Avian Biology* 36: 57–63.

Guichard, Bohoua Louis. 2008. Effect of feather meal feeding on the body weight and feather development of broilers. *European Journal of Scientific Research* 24: 404–409.

Gunderson, Alex R. 2008. Feather-degrading bacteria: A new frontier in avian and host-parasite research? *Auk* 125: 972–979.

Haemig, Paul D. 1978. Aztec emperor Auitzotl and the Great-Tailed Grackle. *Biotropica* 10: 11–17.

———. 1979. The secret of the Painted Jay. *Biotropica* 11: 81–87.

Hansell, Michael H. 2000. *Bird nests and construction behaviour*. Cambridge: Cambridge University Press.

Harpole, Tom. 2005. Falling with the falcon. *Air and Space Magazine*. http://www.airspacemag.com/flight-today/falcon.html. Accessed August 3, 2010.

Harrison, Hal H. 1979. *A field guide to western bird nests*. New York: Houghton Mifflin.

Hart, Ivor B. 1963. *The mechanical investigations of Leonardo da Vinci*. Berkeley and Los Angeles: University of California Press.

Harter, Jim. 1979. *Animals: 1419 copyright-free illustrations of mammals, birds, fish, insects, etc*. New York: Dover Publications.

Hayden, Carl. 1913. Speech: The ostrich industry. In *Congressional Record: Proceedings and Debates of the 62nd Congress, 3rd Session* 49, no. 5: 57–61.

Hecht, M. K., J. H. Ostrom, G. Viohl, and P. Wellnhofer, eds. 1985. *The beginnings of birds: Proceedings of the International "Archaeopteryx" Conference, Eichstatt, 1984*. Willibaldsburg, Germany: Freunde des Jura-Museums Eichstatt.

Hedenström, A., L. C. Johansson, and G. R. Spedding. 2009. Bird or bat: Comparing airframe design and flight performance. *Bioinspiration and Biomimetics* 4: 1–13.

Hedenström, A., L. C. Johansson, M. Wolf, R. von Busse, Y. Winter, and G. R. Spedding. 2007. Bat flight generates complex aerodynamic tracks. *Science* 316: 894–897.

Heilmann, Gerhard. 1927. *The origin of birds*. New York: D. Appleton.

Heinrich, Bernd. 2003a. Overnighting of Golden-crowned Kinglets during winter. *Wilson Bulletin* 115: 113–114.

———. 2003b. *Winter world: The ingenuity of animal survival*. New York: Ecco.

Hill, G. E. 2010. *National Geographic bird coloration*. Washington, DC: National Geographic.

Hill, G. E., and K. J. McGraw, eds. 2006a. *Bird coloration*. Vol. 1, *Mechanisms and measurements*. Cambridge: Harvard University Press.

———. 2006b. *Bird coloration*. Vol. 2, *Function and evolution*. Cambridge: Harvard University Press.

Hingee, Mae, and Robert D. Magrath. 2009. Flights of fear: A mechanical wing whistle sounds the alarm in a flocking bird. *Proceedings of the Royal Society B* 276: 4173–4179.

History of writing materials: The history of the quill pen. 1838. *Saturday Magazine*, January 13, 14–16.

Hornaday, William T. 1913. Woman, the juggernaut of the bird world. *New York Times*, February 23, X1.

Houlihan, Patrick F. 1986. *The birds of ancient Egypt*. Warminster, England: Aris and Phillips.

Houston, David C. 2010a. The impact of red feather currency on the population of the Scarlet Honeyeater on Santa Cruz. In *Ethno-ornithology: Birds, indigenous people, culture, and society*, ed. Sonia Tidemann and Andrew Gosler, 55–66. London: Earthscan.

———. 2010b. The Maori and the Huia. In *Ethno-ornithology: Birds, indigenous people, culture, and society*, ed. Sonia Tidemann and Andrew Gosler, 49–54. London: Earthscan.

Howell, Thomas R., and George A. Bartholomew. 1962. Temperature regulation in the Red-tailed Tropic Bird and the Red-footed Booby. *Condor* 64: 6–18.

How steel pens are made. 1857. *United States Magazine* 4, no. 1: 348–356.

Hu, Dongyu, Lianhai Hou, Lijung Zhang, and Xing Xu. 2009. A pre-*Archaeopteryx* troodontid theropod from China with long feathers on the metatarsus. *Nature* 461: 640–643.

Huxley, Thomas H. 1868. On the animals which are most nearly intermediate between birds and reptiles. *Popular Science Review* 7: 237–247.

———. 1870. Further evidence of the affinity between the dinosaurian reptiles and birds. *Quarterly Journal of the Geological Society of London* 26: 12–31.

Illustrations of cheapness: The steel pen. 1850. *Household Words* 1, no. 24 (1850): 553–555.

Ingham, Phillip W., and Marysia Placzek. 2006. Orchestrating ontogenesis: Variations on a theme by Sonic Hedgehog. *Nature Reviews: Genetics* 7: 841–850.

Ives, Paul P. 1938. *The American standard of perfection*. St. Paul, MN: American Poultry Association.

Jack, Anthony. 1953. *Feathered wings: A study of the flight of birds*. London: Methuen.

Johnsgard, Paul A. 1994. *Arena birds: Sexual selection and behavior*. Washington, DC: Smithsonian Institution Press.

Jones, Eric M. 1996. Hammer and feather. In *Apollo 15 lunar surface journal*. http://www.hq.nasa.gov/alsj/a15/a15.clsout3.html. Accessed July 13, 2010.

Jones, Terry D., et al. 2000. Nonavian feathers in a late Triassic archosaur. *Science* 288: 2202–2205.

Jovani, Roger, and David Serrano. 2001. Feather mites (Astigmata) avoid moulting wing feathers of passerine birds. *Animal Behaviour* 62: 723–727.

Kelson, George M. 1895. *The salmon fly: How to dress it and how to use it*. London: Wyman and Sons.

Kondamudi, Narasimharao, Jason Strull, Mano Misra, and Susanta K. Mohapatra. 2009. A green process for producing biodiesel from feather meal. *Journal of Agricultural and Food Chemistry* 57: 6163–6166.

Laburn, Helen P., and D. Mitchell. 1975. Evaporative cooling as a thermoregulatory mechanism in the fruit bat, *Rousettus aegyptiacus*. *Physiological Zoology* 48: 195–202.

LeBaron, Geoffrey. 2009. The 109th Christmas bird count. *American Birds* 63: 2–9. http://www.audubon.org/bird/cbc.

Li, Quanguo, Ke-Qin Gao, Jakob Vinther, Matthew D. Shawkey, Julia A. Clarke, Liliana D'Alba, Qingjin Meng, Derek E. G. Briggs, and Richard O. Prum. 2010. Plumage color patterns of an extinct dinosaur. *Science* 327: 1369–1372.

Lilienthal, Otto. 2001. *Birdflight as the basis of aviation.* 1889. Reprint, American Hummelstown, PA: Aeronautical Archives.

Lingham-Soliar, Theagarten, Alan Feduccia, and Xiaolin Wang. 2007. A new Chinese specimen indicates that "protofeathers" in the Early Cretaceous theropod dinosaur *Sinosauropteryx* are degraded collagen fibres. *Proceedings of the Royal Society B* 274: 1823–1829.

Lombardo, Michael P., Ruth M. Bosman, Christine A. Faro, Stephen G. Houtteman, and Timothy S. Kluisza. 1995. Effect of feathers as nest insulation on incubation behavior and reproductive performance of Tree Swallows (*Tachycineta bicolor*). *Auk* 112: 973–981.

Long, John, and Peter Schouten. 2008. *Feathered dinosaurs: The origin of birds*. Oxford: Oxford University Press.

Longrich, Nick. 2006. Structure and function of hindlimb feathers in *Archaeopteryx lithographica*. *Paleobiology* 32: 417–431.

Lucas, Alfred M., and Peter R. Stettenheim. 1972. *Avian anatomy—integument.* Washington, DC: U.S. Department of Agriculture.

Lyver, P. O'B., and H. Moller. 1999. Modern technology and customary use of wildlife: The harvest of Sooty Shearwaters by Rakiura Maori as a case study. *Environmental Conservation* 26: 280–288.

Maderson, Paul F. A., Willem J. Hillenius, Uwe Hiller, and Carla C. Dove. 2009. Toward a comprehensive model of feather regeneration. *Journal of Morphology* 270: 1166–1208.

Marchand, Peter J. 1996. *Life in the cold: An introduction to winter ecology*. Hanover, NH: University Press of New England.

Martineau, Lucie, and Jacques Larochelle. 1988. The cooling power of pigeon legs. *Journal of Experimental Biology* 136: 193–208.

Mather, Monica H., and Raleigh J. Robertson. 1992. Honest advertisement in flight displays of bobolinks (*Dolychonyx oryzivorus*). *Auk* 109: 869–873.

Mayr, Gerald, Burkhard Pohl, and Stefan Peters. 2005. A well-preserved *Archaeopteryx* specimen with theropod features. *Science* 310: 1483–1486.

McFarland, Marvin W., ed. 1953. *The papers of Wilbur and Orville Wright.* New York: McGraw-Hill.

McGovern, Victoria. 2000. Recycling poultry feathers: More bang for the cluck. *Environmental Health Perspectives* 108: A366–A369.

Moller, Anders Pape. 1984. On the use of feathers in birds' nests: Predictions and tests. *Ornis Scandivacia* 15: 38–42.

Morgan, Edwin. 1996. *Collected poems*. Manchester: Carcanet Press.

Mynott, Jeremy. 2009. *Birdscapes: Birds in our imagination and experience*. Princeton: Princeton University Press.

Nathan, Leonard. 1998. *The diary of a left-handed birdwatcher*. New York: Harcourt, Brace.

Nelson, Cherilyn N., and Norman W. Henry, eds. 2000. *Performance of protective clothing: Issues and priorities for the 21st century*. Chelsea, MI: American Society for Testing and Materials.

Neme, Laurel. 2009. *Animal investigators: How the world's first wildlife forensics lab is solving crimes and saving endangered species*. New York: Scribner.

Nicholson, Shirley, ed. 1987. *Shamanism*. Wheaton, IL: Quest Books.

Nixon, Rob. 1999. *Dreambirds: The strange history of the ostrich in fashion, food, and fortune*. New York: Picador USA.

Norell, Mark. 2006. *Unearthing the dragon: The great feathered dinosaur discovery*. New York: Pi Press.

Ober, Frederick A. 1905. *Hernando Cortés, conqueror of Mexico*. New York: Harper and Brothers.

Ortega, Francisco, Fernando Escaso, and José L. Sanz. 2010. A bizarre, humped Carcharodontosauria (Theropoda) from the Lower Cretaceous of Spain. *Nature* 467: 203–206.

Ortega-Jiminez, Victor M., and Saul Alvarez-Borrego. 2010. Alcid feathers wet on one side impede air outflow without compromising resistance to water penetration. *Condor* 112: 172–176.

Ostrich "mystery": The solution—Mr. Thornton interviewed. 1911. *Cape Times*, September 27.

Ostrom, John H. 1976. *Archaeopteryx* and the origin of birds. *Biological Journal of the Linnean Society* 8: 91–182.

———. 1979. Bird flight: How did it begin? *American Scientist* 67: 46–56.

Owen, Richard. 1863. On the *Archaeopteryx* of von Meyer with the description of the fossil remains of a long-tailed species, from the lithographic stone of Solnhofen. *Philosophical Transactions of the Royal Society of London* 153: 33–47.

Padian, Kevin. 1983. A functional analysis of flying and walking in pterosaurs. *Paleobiology* 9: 218–239.

———. 1997. A question of emotional baggage. *BioScience* 47: 724.

———. 2001. Cross-testing adaptive hypotheses: Phylogenetic analysis and the origin of bird flight. *American Zoologist* 41: 598–607.

Padian, Kevin, and Kenneth P. Dial. 2005. Could the "four winged" dinosaurs fly? *Nature* 438: E3.

Pagden, Anthony, ed. 2001. *Hernán Cortés: Letters from Mexico*. New Haven: Yale University Press.

Parfitt, Alex R., and Julian F. V. Vincent. 2005. Drag reduction in a swimming Humboldt Penguin, *Spheniscus humboldti*, when the boundary layer is turbulent. *Journal of Bionics Engineering* 2: 57–62.

Pearson, Gilbert T., ed. 1936. *Birds of America*. New York: Doubleday.

Perrichot, V., L. Marion, D. Néraudeau, R. Vullo, and P. Tafforeau. 2008. The early evolution of feathers: Fossil evidence from Cretaceous amber of France. *Proceedings of the Royal Society B* 275: 1197–1202.

Peters, Winfried S., and Dieter Stefan Peters. 2009. Life history, sexual dimorphism, and "ornamental" feathers in the Mesozoic bird *Confuciusornis sanctus*. *Biology Letters* 5: 817–820.

Petroski, Henry. 2007. *The toothpick*. New York: Alfred A. Knopf.

Piersma, Theunis, and Mennobart R. Van Eerden. 1988. Feather-eating in Great Crested Grebes *Podiceps cristatus*: A unique solution to the problems of debris and gastric parasites in fish-eating birds. *Ibis* 131: 477–486.

Pollard, John. 1977. *Birds in Greek life and myth*. London: Thames and Hudson.

Poole A. J., J. S. Church, and M. G. Huson. 2009. Environmentally sustainable fibers from regenerated protein. *Biomacromolecules* 10: 1–8.

Poopathi, Subbiah, and S. Abidha. 2008. Biodegradation of poultry waste for the production of mosquitocidal toxins. *International Biodeterioration and Biodegradation* 62: 479–482.

Pope, Saxton. 1918. Yahi archery. *University of California Publications in American Archaeology and Ethnology* 13, no. 3: 103–152.

———. 1925. *Hunting with the bow and arrow*. New York: G. P. Putnam and Sons.

Price, Jennifer. 1999. *Flight maps: Adventures with nature in modern America*. New York: Basic Books.

Prum, Richard O. 1999. Development and evolutionary origin of feathers. *Journal of Experimental Zoology* 285: 291–306.

———. 2002. Why ornithologists should care about the theropod origin of birds. *Auk* 119: 1–17.

———. 2005. Evolution of the morphological innovations of feathers. *Journal of Experimental Zoology* 304B: 570–579.

———. 2008a. Leonardo da Vinci and the science of bird flight. In *Leonardo da Vinci: Drawings from the Biblioteca Reale in Turin*, ed. Jeannine A. O'Grody, 111–117. Birmingham: Birmingham Museum of Art.

———. 2008b. Who's your daddy? *Science* 322: 1799–1800.

Prum, Richard O., and Alan H. Brush. 2002. The evolutionary origin and diversification of feathers. *Quarterly Review of Biology* 77: 261–295.

———. 2003. The origin and evolution of feathers. *Scientific American*. March: 60–69.

Radcliffe, William. 1921. *Fishing from the earliest times*. London: John Murray.

Reeder, William G., and Raymond B. Cowles. 1951. Aspects of thermoregulation in bats. *Journal of Mammalogy* 32: 389–403.

Regal, Philip J. 1975. The evolutionary origin of feathers. *Quarterly Review of Biology* 50: 35–66.

Reid, James W. 1986. *Textile masterpieces of ancient Peru*. New York: Dover.

———. 2005. *Magic feathers: Textile art from ancient Peru*. London: Textile and Art Publications.

Reina, Ruben E., and Kenneth M. Kensinger, eds. 1991. *The gift of birds: Featherwork of native South American peoples*. Philadelphia: University of Pennsylvania Museum of Archaeology and Anthropology.

Revis, Hannah C., and Deborah A. Waller. 2004. Bactericidal and fungicidal activity of ant chemicals on feather parasites: An evaluation of anting behavior as a method of self-medication in songbirds. *Auk* 121: 1262–1268.

Ribak, Gal, Daniel Weihs, and Zeev Arad. 2005. Water retention in the plumage of diving great cormorants *Phalacrocorax carbo sinensis*. *Journal of Avian Biology* 36: 89–95.

Rombauer, Irma, and Marian Rombauer Becker. 1975. *The joy of cooking*. Indianapolis: Bobbs-Merrill.

Roth, Harald H., and Günter Merz, eds. 1997. *Wildlife resources: A global account of economic use*. Berlin: Springer.

Roush, Chris, and Petyr Beck. 2006. *A good night's sleep: The Pacific Coast Feather story*. Seattle: Documentary Media.

Ruspoli, M. 1986. *The cave of Lascaux: The final photographs*. New York: Harry N. Abrams.

Sahagun, Bernadino de. 1963. *Florentine Codex: General history of the things of New Spain*. Bk. 11, *Earthly things*. Trans. C. E. Dibble and A. J. O. Anderson. 1577. Reprint, Santa Fe: University of Utah and School of American Research.

Schimmel, Annemarie. 1993. *The triumphal sun: A study of the works of Jalaloddinn Rumi*. Albany: State University of New York Press.

Schmookler, Paul, and Ingrid V. Sils. 1999. *Forgotten flies*. Millis, MA: Complete Sportsman.

Sellers, Robin M. 1995. Wing-spreading behavior of the cormorant *Phalacrocorax carbo*. *Ardea* 83: 27–36.

Sereno, P. C., R. N. Martinez, J. A. Wilson, D. J. Varricchio, O. A. Alcober, et al. 2008. Evidence for avian intrathoracic air sacs in a new predatory dinosaur from Argentina. *PLoS ONE* 3, no. 9: e3303. doi:10.1371/journal.pone.0003303.

Shipman, Pat. 1998. *Taking wing: "Archaeopteryx" and the evolution of bird flight*. New York: Simon and Schuster.

Smit, D. van Zyl. 1984. Russel Thornton's ostrich expedition to the Sahara, 1911–1912. *Karoo Agric* 3, no. 3: 19–27.

Smith, Frank C. [Various]. Private correspondence with George Aschman, editor of *Cape Times*. Cataloged at CP Nel Museum, Outdshoorn, South Africa.

Stein, Sarah Abrevaya. 2008. *Plumes: Ostrich feathers, Jews, and a lost world of global commerce*. New Haven: Yale University Press.

Stettenheim, Peter H. 2000. The integumentary morphology of modern birds: An overview. *American Zoologist* 40: 461–477.

Strange, Ian J. 1980. The Thin-billed Prion, *Pachyptila belcheri*, at New Island, Falkland Islands. *Le Gerfaut* 70: 411–445.

Swadling, Pamela. 1996. *Plumes from paradise*. Boroko: Papua New Guinea National Museum.

Tattersall, Glenn J., Denis V. Andrade, and S. Abe Augusto. 2009. Heat exchange from the toucan bill reveals a controllable vascular thermal radiator. *Science* 325: 468–470.

Templeton, Christopher N., and Erick Greene. 2007. Nuthatches eavesdrop on variations in heterospecific chickadee mobbing alarm calls. *Proceedings of the National Academy of Sciences* 104: 5479–5482.

Templeton, Christopher N., Erick Greene, and Kate Davis. 2005. Allometry of alarm calls: Black-Capped Chickadees encode information about predator size. *Science* 308: 1934–1937.

Thaler, Ellen. 1990. *Die Goldhähnchen*. Wittenburg Lutherstadt, Germany: A. Ziemsen Verlag.

Tian, Xiaodong, Jose Iriarte-Diaz, Kevin Galvao Middleton, Israeli Ricardo, Emily Israeli, Abigail Roemer, Allyce Sullivan, Arnold Song, Sharon Swartz, and Kenneth Breuer. 2006. Direct measurements of the kinematics and dynamics of bat flight. *Bioinspiration and Biomimetics* 1: S10–S18.

Tieleman, B. Irene, and Joseph B. Williams. 1999. The role of hyperthermia in the water economy of desert birds. *Physiological and Biochemical Zoology* 72: 87–100.

Tobalske, Bret W. 2007. Biomechanics of bird flight. *Journal of Experimental Biology* 210: 3135–3146.

Tobin, James. 2003. *To conquer the air: The Wright brothers and the great race for flight*. New York: Free Press.

Torre-Bueno, Jose R. 1978. Evaporative cooling and water balance during flight in birds. *Journal of Experimental Biology* 75: 231–236.

Tucker, Vance A. 1968. Respiratory exchange and evaporative water loss in the flying Budgerigar. *Journal of Experimental Biology* 48: 67–87.

Turner, A. H., P. J. Makovicky, and M. A. Norell. 2007. Feather quill knobs in the dinosaur *Velociraptor*. *Science* 317: 1721.

Vuilleumier, François. 2005. Dean of American ornithologists: The multiple legacies of Frank M. Chapman of the American Museum of Natural History. *Auk* 122: 389–402.

Wallace, Alfred Russel. 1869. *The Malay Archipelago*. New York: Harper and Brothers.

Walton, Izaak. 1896. *The compleat angler; or, The contemplative man's recreation*. 1676. Reprint, London: J. M. Dent.

Ward, Jennifer, Dominic J. McCafferty, David C. Houston, and Graeme D. Ruxton. 2008. Why do vultures have bald heads? The role of postural adjustment and bare skin areas in thermoregulation. *Journal of Thermal Biology* 33: 168–173.

Ward, S., U. Möller, J. M. V. Rayner, D. M. Jackson, D. Bilo, W. Nachtigall, and J. R. Speakman. 2001. Metabolic power, mechanical power, and efficiency during wind tunnel flight by European starlings *Sturnus vulgaris*. *Journal of Experimental Biology* 204: 3311–3322.

Ward, S., J. M. V. Rayner, U. Möller, D. M. Jackson, W. Nachtigall, and J. R. Speakman. 2002. Heat transfer from starlings *Sturnus vulgaris* during flight. *Journal of Experimental Biology* 202: 1589–1602.

Wead, E. Young, 1911. The feather industry. *Hunter, Trader, Trapper* 22, no. 5: 23–26.

Winkler, David W. 1993. Use and importance of feathers as nest lining in Tree Swallows (*Tachycineta bicolor*). *Auk* 110: 29–36.

Witmer, Mark. 1996. Consequences of an alien shrub on the plumage coloration and ecology of Cedar Waxwings. *Auk* 113: 735–743.

Wolf, Blair O., and Glenn E. Walsberg. 2000. The role of the plumage in heat transfer processes of birds. *American Zoologist* 40: 575–584.

Xu, Xing, James M. Clark, Jinyou Mo, Jonah Choiniere, Catherine A. Forster, et al. 2009. A Jurassic ceratosaur from China helps clarify avian digital homologies. *Nature* 459: 940–944.

Xu, Xing, Z.-L. Tang, and X.-L. Wang. 1999. A therizinosaurid dinosaur with integumentary structures from China. *Nature* 399: 380–384.

Xu, Xing, X.-L. Wang, and Xiaocun Wu. 1999. A dromaeosaurid dinosaur with a filamentous integument from the Yixian Formation of China. *Nature* 401: 262–265.

Xu, Xing, Xiaoting Zheng, and Hailu You. 2009. A new feather type in a nonavian theropod. *Proceedings of the National Academy of Sciences* 106: 832–834.

———. 2010. Exceptional dinosaur fossils show ontogenetic development of early feathers. *Nature* 464: 1338–1341.

Xu, Xing, Z. Zhou, and X. Wang. 2000. The smallest known non-avian theropod dinosaur. *Nature* 408: 705–708.

Yang, Shu-hui, Yan-chun Xu, and Da-wei Zhang. 2006. Morphological basis for waterproof characteristic of bird plumage. *Journal of Forestry Research* 17: 163–166.

Yanoviak, Stephen P., Robert Dudley, and Michael Kaspar. 2005. Directed aerial descent in canopy ants. *Nature* 433: 624–626.

Zhang, Fucheng, Stuart L. Kearns, Patrick J. Orr, Michael J. Benton, Zhonghe Zhou, Diane Johnson, Xing Xu, and Xiaolin Wang. 2010. Fossilized melanosomes and the colour of Cretaceous dinosaurs and birds. *Nature* 463: 1075–1078.

Zheng, Xiao-Ting, Hai-Lu You, Xing Xu, and Zhi-Ming Dong. 2009. An Early Cretaceous heterodontosaurid dinosaur with filamentous integumentary structures. *Nature* 458: 333–336.

Image and Quotation Credits

A wide range of individuals, publishers, and institutions granted permission for the use of images and quotations in this book. They are gratefully acknowledged here.

xvi Vulture and Book. Painting © 1990 by Simon Thomsett. Used by permission.

1 Introduction epigraph from *The Diary of a Left-handed Birdwatcher* by Leonard Nathan (1996). Used by permission of Graywolf Press.

3 American Robins. Painting from *The Birds of America* by John James Audubon (1840). Reproduction © 2008 by Dover Publications, Inc. Used by permission.

6 Chauvet Owl. Photo by HTO, from Wikimedia Commons (public domain).

8 In-text quotation reprinted with the permission of Simon & Schuster, Inc., from *One River: Explorations and Discoveries in the Amazon Rain Forest*, by Wade Davis. Copyright © 1996 by Wade Davis. All rights reserved.

9 Seraphim Mosaic. Photo by Mattana, from Wikimedia Commons (public domain).

13 Chapter epigraph from *The Archaeopteryx's Song* by Edwin Morgan (1977). Used by permission of Carcanet Press.
18 Cartoon of Charles Darwin from the *Hornet*, March 22, 1871. Image from Wikimedia Commons (public domain).
20 *Archaeopteryx lithographica* (cast of Berlin Specimen). Photo © 2010 by Thor Hanson.
31 Insect Scoop. Illustration by John Ostrom, from Ostrom 1979. Courtesy of *American Scientist*.
35 Winter Wren, woodcut by Thomas Bewick (eighteenth century). Reproduction © 2004 by Dover Publications, Inc. Used by permission.
37 Contour Feather. Illustration © 2010 by Nicholas Judson. Used by permission.
40 Developmental Theory (after Prum 1999). Illustration © 2010 by Nicholas Judson. Used by permission.
43 Chapter epigraph from *Unearthing the Dragon* (2005) by Mark Norell. Used by permission.
49 *Sinosauropteryx prima*. Artwork © 2006 by Julius Csotonyi. Used by permission.
50 *Caudipteryx zoui*. Illustration © 2008 by Peter Schouten. Used by permission.
51 *Beipiaosaurus*. Illustration © by Xing Lida and Zhao Chuang. Used by permission.
60 Common Pigeon. Artist unknown (nineteenth-century engraving). Reproduction © 1979 by Dover Publications, Inc. Used by permission.
63 Thin-billed Prion Chick. Photo © 2005 by Petra Quillfeldt. Used by permission.
68 Dürer's Rhinoceros. Illustration by Albrecht Dürer, 1515. Image from Wikimedia Commons (public domain).
76 Feather Growth and Molt (after Gill 2007). Illustration © 2010 by Nicholas Judson. Used by permission.
77 King of Saxony Bird of Paradise, from Frith and Beehler 1998. Illustrations by William T. Cooper, © 1998 by Oxford University Press. Used by permission.
89 Golden-crowned Kinglet. Illustration by R. Bruce Horsfall, from Pearson 1936 (public domain).

97 Feather Factory. Photo © 2010 by Thor Hanson.

101 Chapter epigraph reproduced from *Analysis of Respiratory Pattern During Panting in Fowl* by Mike Gleeson (1985). Used by permission of the *Journal of Experimental Biology*.

104 Northern Flickers. Painting from *The Birds of America* by John James Audubon (1840). Reproduction © 2008 by Dover Publications, Inc. Used by permission.

107 Girdle-tailed Lizard. Artist unknown (nineteenth century). Reproduction © 1979 by Dover Publications, Inc. Used by permission.

108 Sooty Tern. Artist unknown, from Drent 1972. Used by permission of Koninklijke Brill NV.

110 Leghorn Rooster Feather Tracts. Artist unknown, from Lucas and Stettenheim 1972 (public domain).

112 Toucans. Thermal imagery © by Glenn Tattersall. Used by permission.

114 Bat and Owl. Thermal imagery © Hristov, Allen, and Kunz, Boston University. Used by permission.

115 Flight section epigraph from *Life, the Universe, and Everything* by Douglas Adams, copyright © 1982 by Serious Productions Ltd. Used by permission of Crown Publishers, a division of Random House, Inc.

119 Silver-laced Wyandotte by A. O. Schilling, from *The American Standard of Perfection* (1938 edition). Used by permission of the American Poultry Association.

122 Proavis. Illustration from *Origin of Birds* by Gerhard Heilmann (1927) (public domain).

124 Wallace's Flying Frog by John Gerrard Keulemans, from *The Malay Archipelago* by Alfred Russel Wallace (1869) (public domain).

128 Chukar and Protobird. Artwork by Robert Petty. Courtesy of the Flight Laboratory, Division of Biological Sciences, University of Montana.

133 Apollo 15 Feather. Photo courtesy of NASA.

137 Peregrine Falcons. Painted by Luis Agassiz Fuertes, from Eaton 1915 (public domain).

143 *The Lament for Icarus*. Artwork by Herbert James Draper (1898). Image from Wikimedia Commons (public domain).

147 Lilienthal Glider. Photo courtesy of Otto Lilienthal Museum, Anklam, Germany.

149 Featherfly. Model design by Ray Malmstrom. Courtesy of Impington Village College Model Aeroplane Club and the Ray Malmstrom family.
153 Cardinal Landing. Photo © 2007 by Howard Cheek. Image from BigStock.com.
162 Greater Birds of Paradise by T. W. Wood, from *The Malay Archipelago* by Alfred Russel Wallace (1869) (public domain).
163 Birds of Paradise, from Frith and Beehler 1998. Illustrations by William T. Cooper, © 1998 by Oxford University Press. Used by permission.
168 Sing-sing. Photo © 1991 by Clifford B. Frith. Used by permission.
172 Las Vegas Showgirls. Used by permission of Found Image Press, LLC.
177 McCall's Covers, from *McCall's Magazine*, various issues, 1908–1911 (public domain).
181 Ostrich Expedition Map (after Smit 1984). Map © 2010 by Nicholas Judson. Used by permission.
184 Ostrich Expedition Photos. Courtesy of Dave Glenister and the family of Russel William Thornton.
194 Leah C. Hat. Photo © by 2010 Leah C. Couture Millinery (Photographer: M. K. Semos; hair: Ryan B. Anthony; makeup: Jules Waldkoetter; model: Lindsay Michelle Nader). Used by permission.
195 Chapter epigraph from *Kodachrome* copyright © 1973 Paul Simon. Used by permission of the publisher, Paul Simon Music.
199 Cedar Waxwing. Illustration by R. Bruce Horsfall, from Pearson 1936 (public domain).
202 *Anchiornis huxleyi*. Artwork © 2009 by Julius Csotonyi. Used by permission.
204 Red Feather Money. Photos by William Davenport. Courtesy of the Penn Museum, image numbers 176014 and 176008a.
207 Aztec Warriors. Illustration from *The Florentine Codex*, by Bernadino de Sahagun (1574). Image from Wikimedia Commons (public domain).
218 Water Drop on Feather. Photo courtesy of Edward Bormashenko and the *Journal of Colloid and Interface Science*.
229 Atlantic Salmon Flies. Artist unknown, from *The Salmon Fly* by George Kelson (1895).

231 Fly Casting. Artist unknown, from *The Salmon Fly* by George Kelson (1895).

235 Art of Writing. Artist unknown, from *L'Encyclopédie* by Denis Diderot (1750–1765). Image courtesy of ARTFL Encyclopédie Project.

240 Sower and the Seed. Illumination by Aidan Hart with contributions from Donald Jackson and Sally Mae Joseph, © 2002, The Saint John's Bible, Hill Museum & Manuscript Library, Order of Saint Benedict, Collegeville, Minnesota, USA. Scripture quotations are from the New Revised Standard Version of the Bible, Catholic Edition, Copyright 1993, 1989 National Council of the Churches of Christ in the United States of America. Used by permission. All rights reserved.

242 Quill Pens. Artist unknown, from *L'Encyclopédie* by Denis Diderot (1750–1765). Image courtesy of ARTFL Encyclopédie Project.

249 White-backed Vultures. Watercolor © 1990 by Simon Thomsett. Used by permission.

254 Club-winged Manakin. Illustration © 1998 by Kimberly Bostwick. Used by permission.

257 Manakin Feathers. Artist unknown, from Darwin 1871 (public domain).

268 Feather Lab Plume. Photo © 2010 by Thor Hanson.

273 Illustrated Guide to Feathers. Illustrations © 2010 by Nicholas Judson. Used by permission.

Index